Stochastic Modelling in Process Technology

This is volume 211 in
MATHEMATICS IN SCIENCE AND ENGINEERING
Edited by C.K. Chui, *Stanford University*

A list of recent titles in this series appears at the end of this volume.

Stochastic Modelling in Process Technology

Herold G. Dehling
DEPARTMENT OF MATHEMATICS
RUHR-UNIVERSITÄT BOCHUM
BOCHUM, GERMANY

Timo Gottschalk
DEPARTMENT OF MATHEMATICS
RUHR-UNIVERSITÄT BOCHUM
BOCHUM, GERMANY

Alex C. Hoffmann
DEPARTMENT OF PHYSICS AND TECHNOLOGY
UNIVERSITY OF BERGEN
BERGEN, NORWAY

ELSEVIER
Amsterdam – Boston – Heidelberg – London – New York – Oxford
Paris – San Diego – San Francisco – Singapore – Sydney – Tokyo

Elsevier
Radarweg 29, PO Box 211, 1000 AE Amsterdam, The Netherlands
Linacre House, Jordan Hill, Oxford OX2 8DP, U.K.

First edition 2007

Library of Congress Cataloging-in-Publication Data
A catalog record for this book is available from the Library of Congress

British Library Cataloguing in Publication Data
A catalogue record for this book is available from the British Library

ISBN: 978-0-444-52026-5
ISSN: 0076-5392

For information on all Elsevier publications
visit our website at books.elsevier.com

Transferred to digital printing 2007

07 08 09 10 11 10 9 8 7 6 5 4 3 2 1

Preface

Traditionally in process technology process models are developed by formulating differential balance equations and solving these, sometimes analytically, possibly after some physically reasonable simplification, and sometimes numerically, such as in the rapidly developing field of computational fluid dynamics. Stochastic modeling, on the other hand, focuses on one particle (or fluid pocket or molecule) in the process, and formulates a process model by computing the probability distribution for the state of this particle as a function of time.

As this book attempts to demonstrate, using the stochastic approach for formulating process models has emerged as an efficient and attractive alternative to traditional modeling in many contexts, even for processes that, due to the law of large numbers, in essence behave deterministically.

We aim at introducing stochastic modeling in such a way as to bridge the—unfortunately widening—gap between the fields of (applied) mathematics and process technology. Much of this gap is simply due to differences in terminology and manner of exposure rather than to differences in substance. As this book illustrates, the same issues are being researched by workers in both fields, often with the same or similar or complementary results, but the awareness of progress in the other field is often lacking. We hope to have found a manner of exposure in this text that is readable for members of both fields and in this way contribute to bridging the gap and promoting collaboration.

A good illustration of the strategy we have adopted for this is a quote from a paper on prime numbers and Brownian motion by Patrick Billingsley [16]:

> ...since for the most part I shall only illustrate general results by examples and special cases. For this there is the authority of William Feller, who used to tell us, his students, that the best in mathematics, as in art, letters and all else—that the best consists of the general embodied in the concrete. Although at first I thought that was simply an antimilitary sentiment, I did

> eventually understand it as the intellectual-esthetic principle he intended and have tried ever since to keep it at the front of my mind.

In the same spirit, we have attempted to introduce the principles behind stochastic modeling in process technology by specific examples that contain the basic elements of each type of modeling technique, allowing the reader to glean the general principles from understanding of the examples.

The scope of the book is modeling of processes that naturally falls within the field of process technology (or chemical enginering). We have therefore not included processes that more naturally fall in other fields, such as chemistry or medicine/biology.

Many of the examples, but by no means all, have been taken from our own work in the field of stochastic process modeling, which to a large extent has been focused on fluidized beds. Fluidized bed processing is known (and dreaded by process modelers) for its complex nature, wherein processes discrete in nature take place on different scales, which may or may not be related to the scale of the process vessel. Fluidized beds—as indeed all processes involving moving solid particles, bubbles or drops—are therefore obvious candidates for stochastic modeling.

A number of fields, which have been living largely separate lives in the research literature, are based on the principles of stochastic modeling. They are brought together here, and the common denominators, where present, pointed out.

We give short account of the the organization of the book.

Chapter 1 introduces the principles behind modeling in process technology in general, and sets out stochastic modeling on this backdrop. The mathematical foundation for stochastic process modeling based on Markov chains is given in Chapter 2. This chapter also introduces some basic concepts used later in the book. Chapter 3 describes some stochastic models for batch fluidized beds and discusses the mathematical foundation for modeling processes where interference between particles needs to be taken into account an issue that is relevant in many processes, not only fluidized beds.

Chapters 4 and 5 discuss the issue of particle residence time distribution (RTD) in single process vessels and in complex processes involving several vessels with continuous in- and outflow. RTD theory forms an important part of stochastic modeling in the research literature. Chapter 6 focuses on stochastic modeling of stirred tank reactors. Network-of-Zones (NoZ) modeling and stochastic modeling of processes with chemical reaction are discussed here. Chapter 7 is concerned with processes where the properties of solid particles (such as the size) are modified. The chapter introduces

Population Balance Modeling (PBM), and discusses its relation to Markov chains.

Chapters 8 and 9 discuss formally two model extensions. Chapter 8 introduces a stochastic modeling for multiphase systems. This general theory is applicable to many of the processes discussed in the earlier chapters, and is here, as an example, applied to bubbling fluidized beds with gulf streaming. Finally, Chapter 9 is concerned with the derivation of a continuous stochastic model as the limit of a discrete one.

We hope that this book will convince process technologists to consider stochastic modeling as an alternative to conventional modeling techniques based on conservation equations and their (numerical) solution. We also hope that the book will convince mathematicians that problems within process technology may throw up many new and interesting research topics, and as a spin-off also new mathematical lines of inquiry. But most of all, we hope that it will encourage more collaboration between the two groups of researchers.

Contents

Chapter 1

Modeling in Process Technology

Process technology embraces a wide range of physical, chemical and nuclear processes. modeling within process technology is therefore a complex field, cross-disciplinary in nature and making use of a wide spectrum of analytical and experimental techniques. Moreover, the traditional modeling and design tasks are, to some extent, being taken over by fast-developing software, leaving only the more complex issues as challenges for the scientist or engineer.

This chapter will review briefly the issue of modeling in process technology, starting with an overview of traditional modeling based on formulating and solving balance equations. After that, the concept of stochastic modeling will be introduced using an example.

Let us consider which type of processes we are interested in when formulating models in process technology. If we look at Perry's Chemical Engineer's Handbook [101], we see that of the 15 sections dealing with specific types of processes, 10 are focused on processes involving transport, handling and processing of multiphase streams, *i.e.* streams containing combinations of solids (in the form of particles), liquids and gases. Of these 10 sections, 6 deal with processes involving solids in the form of particles. The 5 that are not concerned with multiphase systems are mostly focused on systems involving single-phase fluids.

We can conclude that systems involving fluid and particle dynamics dominate the processing industry and are therefore the most important class of processes for modeling.

The *objectives* of processes vary widely. If the fluids involved are miscible, the process objective may, for example, be to bring about efficient mixing and dissolution, e.g. in a stirred tank. If immiscible phases are involved, an objective could be to bring the phases into intimate contact for transfer of heat or some constituent from one to the other, or for allowing a catalyst on the surface of solid particles to accelerate a process in a liquid or a gaseous reaction mixture. Another process objective can be to separate phases, often by making use of their different physical properties (mostly density) to bring them into relative motion for separation. Finally a process objective may be transport or storage of fluids or solids.

Figure 1.1 shows some process examples that are common throughout the processing industry. The bubbling fluidized bed is used for its good mixing and heat transfer characteristics. Packed towers are used for contacting immiscible fluids intimately, the packing breaking up the dispersed phase to give a large interfacial area between the phases. The stirred tank is often used for processing or for storage, the stirring making sure that suspension of the dispersed phase is maintained. Pneumatic conveying is an example of a transport process. In most of these processes, several different flow regimes are possible.

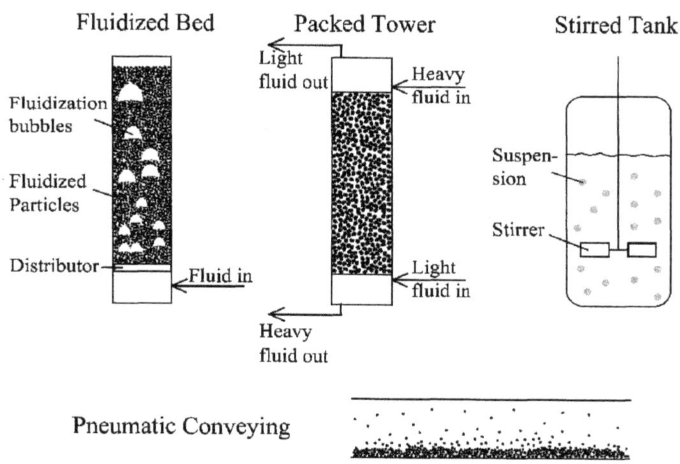

Figure 1.1: Sketches of some processes widely used throughout the processing industry

modeling such processes has traditionally been done by formulating differential balance equations and solving them either analytically or numerically. We shall call this "deterministic modeling". In this book we will discuss an alternative approach: stochastic modeling.

To understand the significance of stochastic modeling and its relation to deterministic modeling we will start this first chapter with a brief review of the latter. We will then introduce stochastic modeling and compare the two approaches for the simple case of one-dimensional convection-diffusion, before going on to more detailed discussions of stochastic modeling in the subsequent chapters.

1.1 Deterministic Modeling

In this section we will review the basic features of deterministic modeling of systems involving fluids, emphasizing aspects that will become relevant in this book. We will introduce the basic concepts by discussing systems where only one phase is present, and will at the end of the section briefly review the way in which the same or similar principles are applied to systems involving more than one phase.

This is not meant to be an exhaustive review of either analytical techniques or research literature, but is intended to give an overview of other modeling techniques currently used in process technology to "set the scene". A few example references, either textbooks or recent publications, where more information about the techniques can be found, will also be given.

In deterministic modeling we define the simplest spatial differential element allowed by the flow configuration and perform a balance over it for some quantity, which is often either the particles or molecules of some constituent in the fluid or heat or momentum. If there are no sources or sinks[1] the balance can be called a "conservation equation", and is:

$$\text{rate of accumulation} = \text{flow in} - \text{flow out}. \tag{1.1}$$

Consider the differential element (or "control volume") sketched in Figure 1.2.

Flows of the quantity we are balancing, for instance heat or the particles or molecules of some constituent, in and out of the element are of two different types:

- *convective transport*, which is transport of the quantity with the flowing fluid, or one could say transport due to organised molecular motion

[1] A source of heat might be heat generation by chemical or nuclear reaction; a source or sink for some chemical constituent may be its production or consumption by chemical reaction

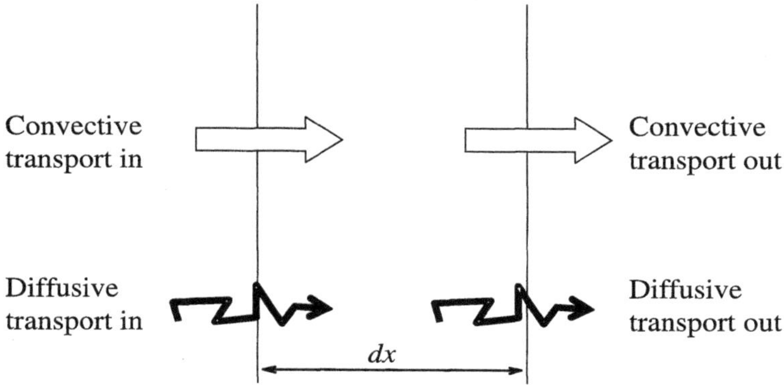

Figure 1.2: Sketch of differential element for one-dimensional balance equation

- *diffusional transport*, which is transport of the quantity due to random molecular motion and molecular interaction.

Convective transport is fairly straight-forward to quantify as we shall see below. To quantify the diffusional in- and outflows three basic laws for diffusional (or molecular) transport are used. They are Fick's law, Fourier's law and Newton's law of viscosity for transport of mass, heat and momentum, respectively. We introduce these laws briefly below.

1.1.1 Basic Laws for Diffusional Transport

We consider uni-directional diffusional transport through an imagined plane (or "control surface") in the fluid. As mentioned, diffusional transport is due to random (or thermal) molecular motion, and is thus in nature a stochastic process. During this motion the molecules will, in a real system, interact in a complex manner. However, the easiest way of picturing diffusional transport on a molecular scale is to consider an ideal system, namely an ideal gas, where the molecules do not interact except as hard spheres to exchange momentum when colliding. The basic laws we will mention are descriptions of the mean behaviour of the system due to this molecular motion. One of the first to relate these laws to molecular motion was Albert Einstein [42].

Note that here and where ever appropriate throughout this book we will base the analyses on *unit cross-sectional area* to keep formulae simple without sacrificing generality. This is equivalent to considering *fluxes* (flows per unit area) rather than flows.

For diffusional transport of *mass*, e.g. the diffusion of particles or molecules of some marked constituent in a mixture of particles or molecules, we use *Fick's law*:

$$J = -D\frac{\partial C}{\partial x}, \tag{1.2}$$

where J is the flux (flow of marked particles or molecules per second per unit area) of the constituent, D is the diffusivity of the constituent in the carrier fluid, C its concentration (number of marked particles or molecules per unit volume) and the x-coordinate is the direction in which diffusion takes place. Note that the definition of D is such that the nett flow of particles or molecules is zero. If it is not, a drift term needs to be added to the right-hand-side of Equation (1.2.

For diffusional transport of *thermal energy*, we use *Fourier's law*:

$$\phi = -k\frac{\partial T}{\partial x} = -\frac{k}{\rho c_p}\frac{\partial(\rho c_p T)}{\partial x}, \tag{1.3}$$

where ϕ is the flux of thermal energy, which is expressed as $\rho c_p T$, where ρ and c_p are the density and the specific heat capacity of the medium, respectively and T is the temperature relative to some datum temperature. k is the thermal conductivity and the x-coordinate is the direction in which heat conduction takes place. The right hand side of the equation shows the flux of thermal energy as a "thermal diffusivity", with the same dimensions as D, length squared over time, L^2/t, multiplying the gradient of the thermal energy per unit volume.

For diffusional transport of *momentum*, the situation is somewhat more complicated. Momentum, $m\mathbf{v}$ per unit volume, is directional, so that we need to consider the diffusion of three components through the plane. We can resolve the momentum in one component normal to the plane through which transport takes place and two tangential to the plane.

Engineers and fluid dynamicists are used to considering the "stresses" acting on imagined planes and control volumes in gases or liquids. These stresses are, in fact, diffusional momentum fluxes *through* the surfaces due to molecular motion.

To see this, consider surfaces of unit area in an ideal gas, Figure 1.3. The left-hand figure depicts a physical wall put into the gas. The pressure that acts on the left surface of this physical wall due to the molecules impacting on it is, by Newton's second law, equal to the rate of change of momentum of the impacting molecules. If the molecules' x-velocity is

v_x, with frequency distribution $f(v_x)$ and there are N molecules per unit volume, the momentum change upon impact of one molecule is $2mv_x$, and the rate of momentum change per second for molecules with velocities in a band dx around v_x is $Nv_x 2mv_x f(v_x)dv_x$. The total rate of momentum change is then:

$$N\frac{1}{2}\int 2mv_x^2 f(v_x)dv_x = Nm\int v_x^2 f(v_x)dv_x = Nm\langle v_x^2\rangle, \tag{1.4}$$

the factor 1/2 because we only integrate over the positive half of the symmetrical function $f(v_x)$, since only molecules with a positive v_x impact on the wall.

Considering now the imagined control surface on the right of Figure 1.3. The total transport of momentum due to molecules passing through the plane in both directions is $Nm\int v_x^2 f(v_x)dv_x$, which is the same as in Equation (1.4.

We thus see that the pressure that would act on a physical surface is the same as the momentum flux across an imagined control surface in the same position. More information can be found, for instance, in Chapman and Cowling[22].

Arguments of a similar nature show that the "stress components" tangential to an imagined plane, *i.e.* the shear stress components, are—at least in an ideal gas—actually momentum fluxes due to molecular motion through the plane in the presence of a macroscopic velocity gradient.

The diffusional flux of momentum with direction *normal* to a control plane is thus equal to the normal stress that would act on a solid wall in that position. For this we use the thermodynamic "pressure", assuming that the normal stress component is isotropic, *i.e.* that it is the same in all directions.

To quantify the two stress components *tangential* to the plane we use *Newton's law of viscosity*:

$$\tau = -\mu\frac{\partial v_x}{\partial y} = -\frac{\mu}{\rho}\frac{\partial(\rho v_x)}{\partial y}, \tag{1.5}$$

where τ is the shear stress, μ is the viscosity coefficient, or just the viscosity, the x-coordinate is the flow direction and diffusion of momentum takes place in the y-direction.

As before, the right-hand-side of the equation shows how the momentum flux can be written as a "momentum diffusivity" (the kinematic viscosity) with the same dimensions as D, L^2/t, multiplying the gradient of the momentum per unit volume.

The three stress components acting on each of the three coordinate planes constitute the nine components of the "stress tensor". To be able

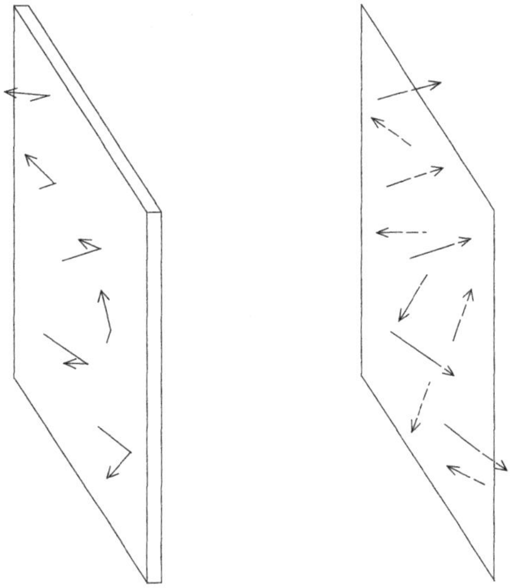

Figure 1.3: Left: the left surface of a physical wall in the fluid experiences a pressure due to the momentum change of the molecules impacting on it as they undergo, in the ideal case, "specular" reflection (equal tangential momentum and equal-and-opposite normal momentum after the reflection). Right: an imagined control volume would experience a numerically equal left-to-right momentum flux due to positive momentum diffusing in the positive direction and negative momentum in the negative direction

to equate the stresses directly to momentum fluxes, one needs to use the sign convention for stresses adopted, among others, by Bird et al.[17].

We note at this point that transport of the nature described by the three equations discussed in this section can be due to random motion other than molecular motion if the fluid in question is flowing. *Turbulent eddies* caused either by friction of the flowing fluid with the conduit walls or by some packing material through which the fluid flows can cause the same type of transport. It is then referred to as "dispersion" rather than diffusion. We note that the dispersion coefficient associated with random motion of fluid elements in this way is, in contrast to the diffusion coefficient associated with random molecular motion, not a material characteristic, but depends also on the flow configuration, and that it may neither be homogeneous in time or space nor isotropic.

1.1.2 The One-Dimensional Convection-Diffusion Equation

We now illustrate the principle of modeling by balance equations by a simple example: deriving the convection-diffusion equation. Consider the simple case of a fluid flowing with a cross-sectionally uniform velocity through a tube with the Cartesian x-coordinate in the flow direction (see Figure 1.4). The tube can be a *stream tube*, which does not have physical walls but is limited by streamlines, such that there is no flow through the boundary. The fluid contains some component with the concentration $C(x,t)$ varying spatially only in the x-direction. For this example and the equivalent stochastic one below, we assume the flow to be laminar, so that the diffusion is due to random molecular motion rather than turbulent eddies, although the presence of turbulent dispersion would not change the analysis, as long as the turbulence is homogeneous.

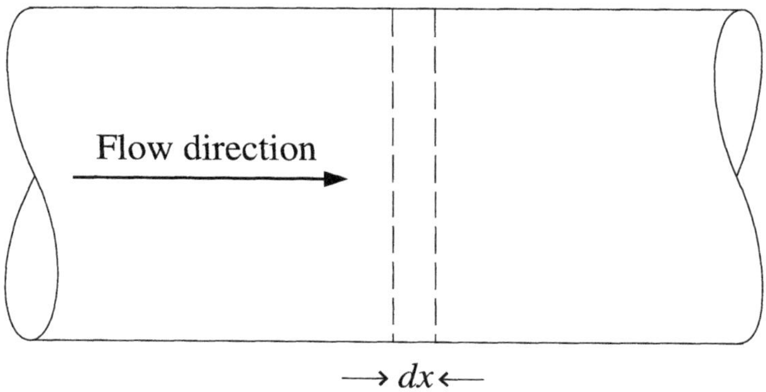

Figure 1.4: Tube section with flowing fluid, the flow direction and the differential element over which the balance equation is written are indicated

Since $C(x,t)$ does not vary in the two other directions, we can define an element that is only differential in the x-direction, as indicated.

In this analysis, and in the rest of this chapter, we will assume diffusivities to be constant in space and time.

Considering, as mentioned before, fluxes rather than flows, or, equivalently, basing the analysis on an x-surface of unit cross-sectional area, we find that the convective influx is simply $v_x C(x,t)$. Thus we obtain for the total influx due to convection and diffusion:

$$J_{in} = \left(v_x C(x,t) - D\frac{\partial C(x,t)}{\partial x} \right)_x \tag{1.6}$$

and for the outflux:

$$J_{out} = \left(v_x C(x,t) - D\frac{\partial C(x,t)}{\partial x} \right)_{x+dx}. \tag{1.7}$$

Writing the outflux in terms of the values at the in-boundary by expanding as a first order Taylor series, assuming v_x and D to be constant gives:

$$J_{out} = v_x \left(C(x,t) + \frac{\partial C(x,t)}{\partial x} dx \right) - D \left(\frac{\partial C(x,t)}{\partial x} + \frac{\partial}{\partial x} \left(\frac{\partial C(x,t)}{\partial x} \right) dx \right). \tag{1.8}$$

The rate of accumulation of the component in the element is:

$$\frac{\partial C(x,t)}{\partial t} \mathrm{d}x. \tag{1.9}$$

Thus, performing the balance indicated by Equation 1.1 and simplifying gives the *one-dimensional convection-diffusion equation*:

$$\frac{\partial C(x,t)}{\partial t} = -v_x \frac{\partial C(x,t)}{\partial x} + D\frac{\partial^2 C(x,t)}{\partial x^2}. \tag{1.10}$$

1.1.3 Other Balance Equations

The type of balance illustrated above is ubiquitous in modeling in process technology, in fact the basic differential equations of change are derived from balances over differential control volume, or elements, in three dimensions.

Deriving these equations, or even giving them in their full form, is beyond the scope of this short review, the reader is referred to the bibliography for more information. We will, however, give the Cartesian form of the continuity equation and the Navier-Stokes equations as examples.

The "equation of continuity" is the equation for mass conservation in a flow field, and can be derived by balancing mass over a differential control volume according to:

rate of accumulation = inflow − outflow

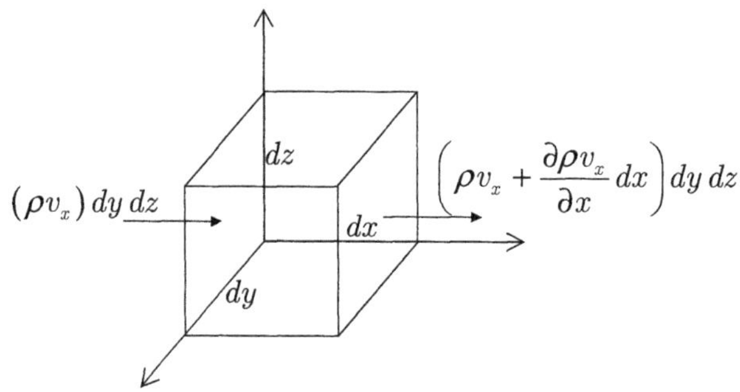

Figure 1.5: Mass flows in the x-direction in and out of a differential control volume in a Cartesian coordinate system

Figure 1.5 shows the principle in a Cartesian coordinate system, where the flows of mass in the x-direction in and out of a differential element are indicated. We have again written the outflow in terms of the inflow by a first-order Taylor expansion.

Setting the sum of the flows in all three coordinate directions equal to the rate of accumulation, $\frac{\partial \rho}{\partial t} dxdydz$ and simplifying results in the equation of continuity in a Cartesian coordinate system:

$$\frac{\partial \rho}{\partial t} + \frac{\partial \rho v_x}{\partial x} + \frac{\partial \rho v_y}{\partial y} + \frac{\partial \rho v_z}{\partial z} = 0. \tag{1.11}$$

The "fluid equations of motion" can be derived in a very similar way by balancing momentum flows over a differential control volume, and relating the forces acting on the volume to the gradients of the velocity field using a generalization of Newton's law of viscosity. In addition to the momentum fluxes arising from flow and diffusion, a source term for momentum arises due to an external force on the fluid, the force of gravity.

These equations of motion reduce to well-known Navier-Stokes equations for an incompressible fluid with constant viscosity. The Cartesian form of the Navier-Stokes equations is:

$$\begin{aligned}
&\rho\left(\frac{\partial v_x}{\partial t}+v_x\frac{\partial v_x}{\partial x}+v_y\frac{\partial v_x}{\partial y}+v_z\frac{\partial v_x}{\partial z}\right)\\
&\qquad=-\frac{\partial p}{\partial x}+\mu\left(\frac{\partial^2 v_x}{\partial x^2}+\frac{\partial^2 v_x}{\partial y^2}+\frac{\partial^2 v_x}{\partial z^2}\right)+\rho g_x\\
&\rho\left(\frac{\partial v_y}{\partial t}+v_x\frac{\partial v_y}{\partial x}+v_y\frac{\partial v_y}{\partial y}+v_z\frac{\partial v_y}{\partial z}\right)\\
&\qquad=-\frac{\partial p}{\partial y}+\mu\left(\frac{\partial^2 v_y}{\partial x^2}+\frac{\partial^2 v_y}{\partial y^2}+\frac{\partial^2 v_y}{\partial z^2}\right)+\rho g_y\\
&\rho\left(\frac{\partial v_z}{\partial t}+v_x\frac{\partial v_z}{\partial x}+v_y\frac{\partial v_z}{\partial y}+v_z\frac{\partial v_z}{\partial z}\right)\\
&\qquad=-\frac{\partial p}{\partial z}+\mu\left(\frac{\partial^2 v_z}{\partial x^2}+\frac{\partial^2 v_z}{\partial y^2}+\frac{\partial^2 v_z}{\partial z^2}\right)+\rho g_z
\end{aligned}\quad. \tag{1.12}$$

Often we are only interested in steady flow, *i.e.* the flow is constant in time, so that the time-derivatives are zero.

Equations (1.12) are in vector form:

$$\rho\left(\frac{\partial \mathbf{v}}{\partial t}+\nabla\cdot\mathbf{v}\right)=-\nabla p+\mu\nabla^2\mathbf{v}+\rho\mathbf{g}. \tag{1.13}$$

Defining dimensionless parameters in terms of a characteristic resultant velocity, v_{ch}, a characteristic length, D_{ch}, the fluid density, ρ, and some arbitrary reference pressure, p_0 gives:

$$\mathbf{v}^*:=\frac{\mathbf{v}}{v_{ch}};\quad p^*:=\frac{(p-p_0)}{\rho v_{ch}^2};\quad t^*:=\frac{t v_{ch}}{D_{ch}};\quad \nabla^*:=D_{ch}\nabla, \tag{1.14}$$

for pipe flow, v_{ch} can, for instance, be the mean fluid velocity and D_{ch} can be the pipe diameter. Inserting these in Equation (1.13) and simplifying gives:

$$\begin{aligned}
&\frac{\partial \mathbf{v}^*}{\partial t^*}+\nabla^*\cdot\mathbf{v}^*=-\nabla^*p^*+\left(\frac{1}{Re}\right)\nabla^{*2}\mathbf{v}^*+\left(\frac{1}{Fr}\right)\frac{\mathbf{g}}{g}\\
&\qquad Re=\frac{\rho v_{ch}D_{ch}}{\mu};\quad Fr=\frac{gD_{ch}}{v_{ch}^2}.
\end{aligned} \tag{1.15}$$

where Re is the Reynolds number, which can be viewed as a measure of the ratio of inertial forces to viscous forces in the flow and Fr is the Froude number, a measure of the ratio of gravitational forces to inertial forces. Re is fundamental in process technology, and its magnitude largely determines the flow pattern in a confined flow and flow around objects. The value of Fr does not affect the flow unless a free, non-horizontal interface between fluids of different densities is present.

These equations, together with the equation of continuity describe, as far as we know at present, all phenomena in the dynamics of single phase incompressible Newtonian fluids, except in situations where the fluid can no longer be considered a continuum. An example where the continuum assumption breaks down is that of the flow of a rarified gas around very small particles or drops, where the effect of the gas consisting of finite molecules cannot be neglected.

Other balance equations are the equations for mechanical energy and total energy. The latter can be applied in a simplified form to a stationary medium, such as a solid, to obtain the diffusion equation for heat (see, for instance, reference [21]). Source terms in the total energy equation can be due to chemical or nuclear reactions. The three-dimensional version of the convection-diffusion equation is a balance equation for some constituent in the flow. Source terms in this equation can be due to production or consumption of the component considered through some chemical or nuclear reaction.

Ample discussion of these and other equations of change, all derived by similar balances, can be found in most text-books on fluid dynamics and transport phenomena, such as Batchelor [5] or Bird et al. [17].

1.1.4 Some Solution Methods

The balance equations are partial differential equations, often strongly coupled vector equations, and they cannot be solved analytically in most practical cases. Much research work in fluid dynamics has always been focused on finding elegant, yet physically relevant, simplifications making analytical solutions of the equations possible, and most textbooks and courses in fluid dynamics and modeling in process technology are devoted to this.

Many of these methods are based on finding symmetry in the flow fields, or in other ways making simplifying assumptions about them *based on the geometry of the flow field.* One example is the laminar flow of an incompressible, Newtonian fluid, which is assumed axisymmetrical with respect to the pipe axis, so that the Navier-Stokes equations can, in a cylindrical coordinate system, be reduced to a simple differential equation for the velocity as a function of the radius, at least if the flow is laminar (which it almost never is in practice).

Simplifying the general equations of change based on the geometry of the system is equivalent to performing a first-principles balance on an optimally configured differential element, as we did to derive the one-

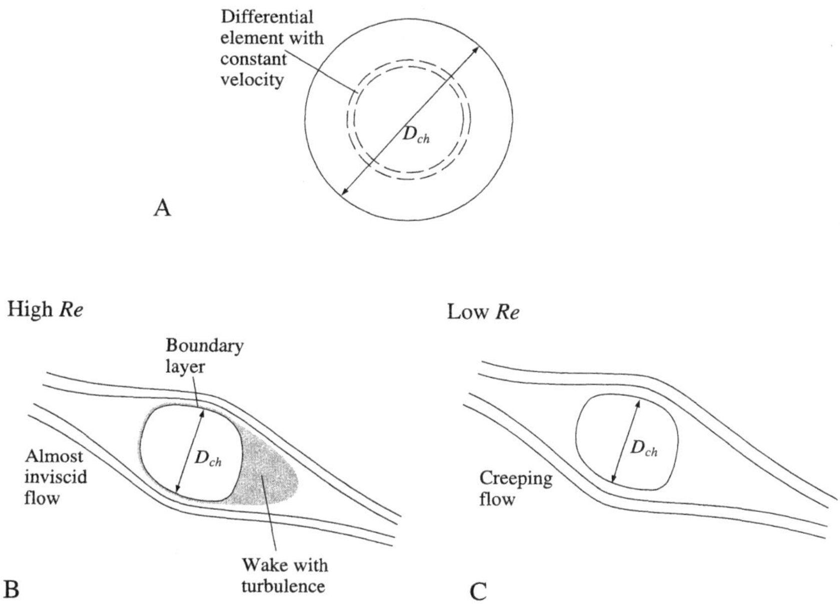

Figure 1.6: Illustrations of simplifications of the Navier-Stokes equations. **A** Cross-sectional view of pipe flow under laminar conditions (low *Re*), which is assumed to be axisymmetric. **B** Flow around a body at high Reynolds number, where the flow outside a narrow boundary layer by the body can be assumed to be almost potential. **C** Creeping flow around a body at low *Re*. Note that some length characteristic for the flow configuration is taken as D_{ch} in *Re*

dimensional convection-diffusion equation above. In Figure 1.6A the principle is illustrated by a sketch of the cross-section of pipe flow at low Reynolds number, $Re = \frac{\rho v D_{ch}}{\mu}$, where the flow is assumed to be axisymmetric, such that the velocity in the differential element shown is constant greatly simplifying the solution of the Navier-Stokes equations for this flow configuration.

Another class of simplifying assumptions, often used in combination with assumptions about the geometry, is to *simplify the equations of motion themselves*, assuming some of the terms to be negligible in comparison with others. One example of this is that at high *Re* the flow around objects can be considered inviscid ("potential flow assumption"), *i.e.* $\mu \to 0$ except for a thin layer around the object, this is illustrated in Figure 1.6B. Thus in the bulk flow outside this thin layer the terms involving the viscosity in the equations of motion can be considered negligible and dispensed with, which simplifies the equations greatly sometimes making them amenable to analytical solution, for instance if the body is spherical. In situations concerning flows of very viscous material of small density and in small confines, the opposite assumption is often made: that viscous forces completely dominate the flow ("creeping flow" see, for instance, reference [54]), i.e. or $Re \to 0$, in which case the "inertial terms", which are the terms involving the velocity components on the right-hand-side of Equations (1.12) and (1.15) can be dispensed with. This is illustrated in Figure 1.6C. Actually the potential and creeping flow patterns look somewhat similar, but the streamlines diverge more around the body in the latter than in the former. Each simplifying assumption obviously has its own field of applicability.

When Re is large enough, turbulent eddies will occur spontaneously in a confined flow, such as pipe flow[2]. Turbulence will be present in most practical cases and is thus ubiquitous in flows in process technology. The effect of turbulence has been modeled in many ways, see for instance references [20, 60, 82]. Turbulence gives rise to random velocity fluctuations in the flow field, and therefore gives rise to extra dispersion of mass, heat and momentum over and above that due to random molecular movement. The simplest way to model the effect of turbulent dispersion of momentum, and therefore the effect of turbulence on the mean velocity field, is to define a "turbulent viscosity", which is added to (and often dwarfs) the molecular viscosity. The turbulent viscosity is, in contrast to the molecular one, not normally homogeneous (the same everywhere in the flow field) or isotropic (the same in all directions). All turbulence models can, in principle, capture the former feature, and the more complex ones the latter.

Analytical solutions to the balance equations, and in particularly to the Navier-Stokes equations, based on simplifying assumptions have grown ever more complicated. Recently, however, a powerful technique has been developed which is beginning to put all other solution techniques in the shadow: *Computational Fluid Dynamics*, or CFD for short. CFD has exploded on to the scene in the last few decades, and is rapidly becoming the all-dominating tool in fluid dynamics research, both in academia and in industry.

In CFD, finite difference equivalents of the differential balance equations are solved on a "computational grid". The computational grid consists of "node points" at which the dependent variable, which can be the temper-

[2]How large Re needs to be for a given flow type to prevail is determined empirically in each flow configuration. For instance, it is found that in pipes turbulent flow prevails for $Re \gtrapprox 10000$, while laminar (or creeping) flow prevails for $Re \lessapprox 2000$. For flow around bodies, the flow pattern is close to creeping for $Re \lessapprox 0.1$

ature, the concentration of some constituent or—as in the Navier-Stokes equations—the momentum per unit volume (density times velocity), is evaluated. A one-dimensional finite difference grid is sketched in Figure 1.7.

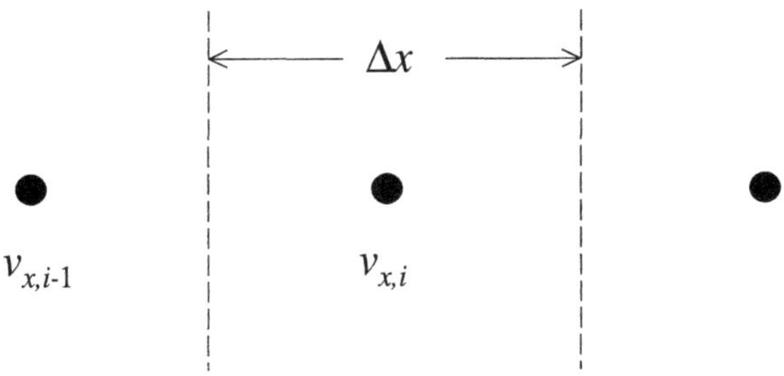

Figure 1.7: Sketch of a computational grid with node points and cell boundaries indicated

The terms in the balance equations are approximated using the values on the computational grid ("finite differencing"). There are different ways of doing this (see, for instance, Patankar [100] or Press et al.[102]), for instance the second term on the left-hand-side of the x-component part of Equation (1.12) can be approximated by:

$$v_{x,i}\frac{v_{x,i+1} - v_{x,i-1}}{2\Delta x} \tag{1.16}$$

—while the second term on the right-hand-side of the same equation can be approximated by:

$$\mu\frac{v_{x,i+1} + v_{x,i-1} - 2v_{x,i}}{\Delta x^2}. \tag{1.17}$$

In this way the balance equations are re-written into finite difference form, and the unknowns of the resulting algebraic equations are the values of the dependent variables on the node-points.

In principle, if the grid is fine enough, this method should yield solutions to the balance equations in any flow configuration, and thus represent the final answer in fluid dynamics. Solving the Navier-Stokes equations on a sufficiently fine grid even results in generation of turbulence in a flowing fluid where appropriate ("direct turbulence simulation").

However, some problems exist. One is the finite difference formulation. The method of finite differencing exemplified by Equations (1.16) and (1.17) is called "central differencing". This may lead to unstable or oscillatory solutions, and to avoid this, "upstream differencing" was devised, emphasizing the upstream node values when formulating expressions for the terms in the balance equations. Upstream differencing, however, can, particularly in cases involving strong cross-grid flow, lead to a phenomenon called "numerical diffusion". This is a diffusion-like effect, which is not related to the physics of the flow, but only to the numerical procedure.

Another problem is that the fine grids necessary for direct turbulence simulation are not yet possible in three dimensions, at least if the finest turbulent eddies are to be simulated correctly, and will not be for some time to come. Meanwhile the effect of turbulence is simulated using "turbulence models" of varying degree of complexity, introducing a significant element of empiricism in the whole process. The simplest turbulence modeling approach is to calculate a turbulent viscosity, as mentioned above, on basis of two balance equations: one for the rate of turbulence generation and one for the rate of its dissipation ("k-ϵ model").

The state-of-the-art, and an exciting development in CFD eliminating many of the assumptions and empirical elements in traditional turbulence modeling, is an intermediate step in the move toward direct turbulence modeling: Large Eddy Simulation (LES), where the large eddies of the turbulence are simulated directly, and a turbulence model (which can be more simple) is used to mimic the effect of the sub-grid turbulence. Derksen, for example, used this technique to simulated the flow in a cyclone separator and in a vortex tube [40, 41].

Introductions to CFD can be found in the excellent, but by now somewhat dated, book of Patankar [100] or in Press et al. [102] or Versteeg et al. [127].

1.1.5 Multiphase Systems

If a mixture consisting of more than one phase is present in the system, one phase will exist as the continuous one, and the others will be dispersed in the form of bubbles, drops or solid particles; we will refer to all three as "particles" in this section.

The dynamics of the continuous phase fluid is governed by the balance equations mentioned above.

For the dispersed phase(s), the transport of mass, thermal energy and momentum inside the particles are also calculated by the balance equations above. If the dispersed particles are fluid, internal flow may be present depending on the relative viscosities of the continuous and dispersed fluids and the presence of surfactants on the interface; the latter may render the interface effectively immobile.

If the full balance equations could be solved—e.g. by CFD—in parallel for the continuous and dispersed phases, the dynamics of each of the phases and the exchange of mass, heat and momentum between the phases would be found directly. In fact, this is not possible due to limitations in computational power, and, to an extent, lack of understanding of the physics governing the interfacial phenomena and the particle-fluid, particle-particle and particle-wall interactions. Other techniques are therefore used.

We briefly review the most important of these techniques, starting with direct calculation but (semi-) empirical methods, and finishing with numerical methods.

By direct calculation the motion of the particles and their effect on the motion of the continuous phase, in other words the exchange of *momentum* between the phases, is usually described by particle equations of motion. These are written in a Lagrangian frame of reference (a "Lagrangian" reference frame is following the particle, as distinct from a "Eulerian" reference frame, which is fixed relative to the coordinate system). This particle equation of motion is simply a statement of Newton's 2. law:

$$m\mathbf{a} = \sum_i \mathbf{F}_i, \tag{1.18}$$

where m and $\mathbf{a}$ are the particles' mass and acceleration and $\mathbf{F}_i$ are the forces acting on it. The full form of this equation is given in Clift et al. [25], together with a wealth of other information about multiphase systems. The forces acting on the particle are flow forces (such as drag, lift and moment) and conservative forces (such as gravity, boyancy, electrostatic or magnetic forces). For solid particles forces and moments may furthermore result from the particles impacting on each other or on solid walls. Reviews of the forces acting on particles in multiphase systems, and the resulting equations of motion, are given in Clift et al. [25] and Crowe et al. [27]. One well-known model for the size of the drag force on small solid spheres, for instance, is Stokes' law:

$$F = 3\pi\mu D_p v_p, \tag{1.19}$$

where v_p is the size of the particle's velocity relative to the fluid.

The interchange of *mass* and *heat* between the phases is often modeled semi-empirically by assuming that the bulk of the continuous phase is well mixed, and a "diffusion" layer exists around the particle through which the molecules or heat has to diffuse in accordance with Equations (1.2) and (1.3), respectively and accounting for the resistance to mass and heat transfer between the continuous and dispersed phases. The thickness of this layer, and therefore the resistance, depends on the relative velocity between particle and continuous phase, and the density and viscosity

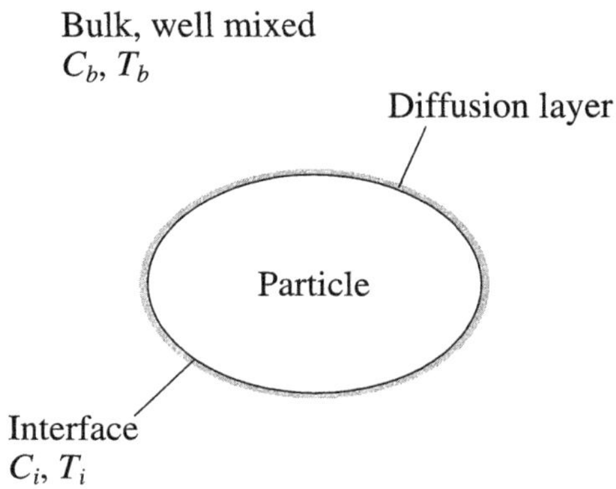

Figure 1.8: Illustration of the principle of modeling mass and heat transfer between the continuous and dispersed phases in a multiphase system

of the continuous phase. Relations for this thickness are mostly empirically determined. The transfer of mass and heat between the bulk of the continuous phase, which is assumed well mixed, and the particles of the dispersed phase are normally assumed to be proportional to the driving forces, namely the concentration and temperature difference, respectively between the bulk and the interface (see Figure 1.8, and the proportionality factors are called the "mass transfer coefficient", k and the "heat transfer coefficient", h. Expressions for the fluxes of mass (molecules) and heat between the phases thus become:

$$\begin{aligned} J_m &= k(C_b - C_i) \\ J_h &= h(T_b - T_i) \end{aligned} \tag{1.20}$$

where the SI-units of J_m are [m s^{-1}], and those of J_h are [J m^{-2} s^{-1} °C^{-1}].

In general one distinguishes between systems in which only the effect of the continuous phase on the dispersed one(s) needs to be considered ("one-way coupling") and those where also the effect of the dispersed phase(s) on the continuous one needs to be accounted for ("two-way coupling"). The former are by far the simplest since the velocity field of the continuous phase can be calculated first disregarding the particles, and the particle paths then calculated using Equation (1.18). Dimensionless, order-of-magnitude, "coupling-parameters" can be defined indicating to which class a given system belongs. For instance, for a gas containing droplets, a coupling-parameter for mass is defined [27] as the mass flow of the gas through a control volume divided by the rate of mass added to, or removed from, the gas in the control volume due to droplet evaporation or condensation.

Reviews of analytical models for the exchange rates of mass, heat and momentum between continuous and dispersed phases, and the significance for modeling reactor performance in process technology are given in references [83, 135]. Such models are often (semi-) empirical.

We now turn to *numerical* solution schemes. As mentioned, numerical solution of the balance equations for each phase in principle makes direct simulation of multiphase systems possible. However, the problems in simulating single phase systems are even worse for multiphase systems. Not only is the structure of such a system very fine due to the presence of the particles, but it is much more complicated to simulate due to the complex interaction between the phases, *i.e.* between the continuous fluid and the particles, between particles themselves and between the particles and the walls of the container.

Nevertheless, simulations are being developed at varying scales. We mention some methods starting with the smallest scale.

Direct, small-scale simulations are being developed, in which the balance equations are solved for the two phases, while the computational grid is periodically adjusted to allow the particles to move under the influence of the forces acting on them. As computing power increases, this will become more important.

For systems where the interaction between solid particles is more important than fluid-solid interaction, for instance for particle beds in storage vessels, rotating driers and mills, "Discrete Element modeling" (DEM) or "granular dynamics" is being developed. This method was initially derived from "Molecular Dynamics Simulations" (MD). Particles are tracked in a Lagrangian manner as they bounce and slide over each other, by solving a version of Equation (1.18), where the forces considered are those due to particle/particle and particle/wall interaction. The motion of the particles, which are normally modeled as spheres, may be modified by including some fluid-particle interaction in the equation of motion, for instance if they move in a liquid. In a recent paper Cleary and Sawley [24] extended the method to non-spherical particles.

Another class of systems is those where the particles interact significantly with fluids, for instance in fluidized beds and in risers, where the particles are supported or transported in up-flowing gas streams. If the particles are dilute in the continuous fluid, one-way coupling [27] can be used in the simulations, so that the continuous phase flow is simulated first, and the particle flow is then simulated using the pre-calculated continuous phase flow field. However, if the particles are so concentrated that they influence the continuous fluid significantly, two-way coupling needs to be taken into account. In this case modified versions of the Navier-Stokes equations are used for the fluid motion: a source-term is included in the equations, which accounts for the momentum interchange with the particles. To quantify the source term, the interaction between the fluid and the dense cloud of particles can be modeled by a semi-empirical relation for the pressure drop due to flow through particle beds, such as the well-known Ergun equation. Also here the particles are tracked in a Lagrangian manner as they move in the fluid and collide with each other and the walls of the vessel (see, for instance, references [18, 70, 129]). this method is called "Eulerian-Lagrangian", since the fluid flow field is calculated in a Eulerian manner, and the particles are tracked in a Lagrangian manner.

Another simulation technique, with which it is easy to model large systems with two-way coupling, is "Eulerian-Eulerian" where the particles are treated as a second "fluid" interpenetrating and interacting with the continuous fluid phase. This method is widely used for modeling particle clouds in gases. A disadvantage of this method that it is less easy to take into account some of the forces acting on the particles, such as lift forces and the effect of collisions.

This ends our brief review of the most popular modeling techniques in process technology, and we will now turn our attention to Stochastic Process modeling, which is the subject of this book.

1.2 Stochastic modeling—an Example

While deterministic modeling is based on formulating and solving differential balance equations, stochastic modeling often focuses on *a single particle* in the system. The word "particle" should be understood in the widest sense, it may be a particle (bubble, drop or solid particle) of a dispersed phase, or it may be en element of fluid or even a molecule in a fluid. A stochastic variable is assigned to the particle property in which we are interested. This can be the position of the particle in a reactor or other vessel, or its residence time in the vessel.

Other types of stochastic models exist, however, where a single particle is not the focus. In population balance modeling, for instance, a stochastic variable is assigned to the number of particles in a given property class where the "property" may be the particle size, composition, density, etc.

We will introduce the basic concepts behind stochastic modeling by means of an example, namely by deriving the one-dimensional convection-diffusion equation, which we derived on basis of balance equations in Section 1.2.1, using a Markov chain model for the particle's position in the vessel.

1.2.1 A Discrete Markov Chain for One-Dimensional Flux

Consider again the case of a fluid flowing with a cross-sectionally uniform velocity through a tube with the Cartesian x-coordinate in the flow direction (see Figure 1.9). The fluid contains some component with the concentration $C(x,t)$ varying only in the x-direction.

We consider the movement of *one molecule of the component in question*. To make the discussion general, we will refer to the molecule as a "particle". We will consider the position of this particle as a function of time. Although it can take up positions anywhere along the tube, we will discretize space to simplify this introduction, so that the particle's position needs only be defined by the cell it occupies, denoted by i. We assume the tube to be finite, so that $1 \leq i \leq N$ (see Figure 1.9). We also discretize time and consider the particle's position at regular time intervals denoted by n with $n \geq 1$.

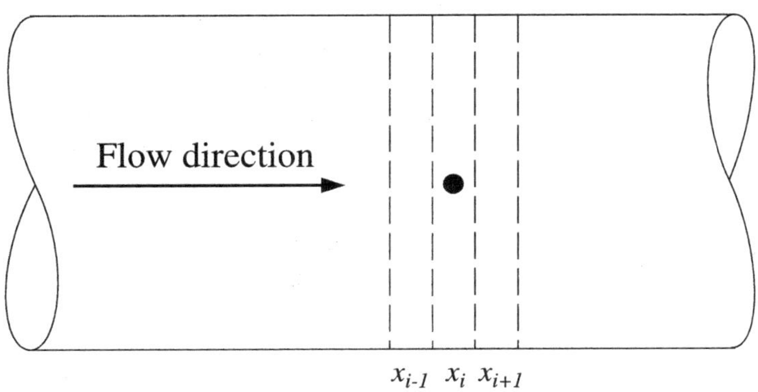

Figure 1.9: Tube section with flowing fluid, the flow direction, the particle considered and the discretization are indicated

In a stochastic approach, we model the position of the particle in the tube at time n by a stochastic ("random") variable: $(X_n)_{n\geq 1}$, X_n is thus the location of the particle at time n. The probability that the particle is in a given location i at time n, which we denote by the vector $\mathbf{p}(n) = p(n,i)_{1\le i\le N; n\ge 0}$, is thus given by the probability distribution of X_n.

An important prerequisite for the type of analysis we will perform is that the process constitutes a *Markov chain i.e.*, that for times $1\dots n+1$, we have:

$$P(X_{n+1} = i_{n+1}|X_1 = i_1,\dots,X_n = i_n) = P(X_{n+1} = i_{n+1}|X_n = i_n), \quad (1.21)$$

which means that the probability of a particle being at position i_{n+1} at some time $n+1$ while having been at position $i_1,\dots,i_n$ at times $1,\dots,n$ before equals the probability of a particle being at i_{n+1} at time $n+1$ while only having been at the point i_n at time n. Thus the probability distribution of the particle's location in the future only depends on the present position i_n and not on the past $i_1,\dots,i_{n-1}$.

The values that X_n can take—in this case $1\le i\le N$—is called the *state space* of the Markov chain.

The particle's *transition probabilities* are defined as $p_{ij}(n) = P(X_n = j|X_{n-1} = i)$ for $1\le i,j\le N; n\ge 0$, or in words: the probability that the particle is in some position j at time n, given that it was in i at time $n-1$. The transition probabilities form an $(N\times N)$ matrix $\mathbf{P}(n)$, and each row sums to unity: $\sum_j p_{ij} = 1$. If the process is *time-homogeneous* (steady), the transition probabilities are independent of n. Specifying the transition probabilities is enough to specify the entire process.

Now let $\mathbf{p}(0) = p(0,i), i = 1\dots N$ denote the probability vector of the particle's initial position, i.e., $P(X_0 = i) = p(0,i)$. Then we can obtain the probability of specific events from the transition probabilities. For instance the probability that the particle starts at time $n = 0$ in cell i_0 and then visits cells $i_1,\dots,i_{n-1}$ to finally end up in cell i_n at time n is: $p(0,i_0)p_{i_0i_1}p_{i_1i_2}\cdots p_{i_{n-1}i_n}$.

In general we can obtain the probability vector at time n, $\mathbf{p}(n)$ from that at the previous time, $n-1$ from the equation:

$$p(n,j) = \sum_{i=1}^{N} p(n-1,i)p_{ij}, \quad (1.22)$$

or in matrix form:

$$\mathbf{p}(n) = \mathbf{p}(n-1)\mathbf{P}(n). \quad (1.23)$$

We can also obtain the total probability that the particle will arrive in cell i_n at time n given that it was in cell i_{n-2} at time $n-2$:

$$P(X_n = i_n|X_{n-2} = i_{n-2}) = p(n-2,i_{n-2})\sum_{i_{n-1}} p_{i_{n-2}i_{n-1}}p_{i_{n-1}i_n}. \quad (1.24)$$

The probability vector at any time n, $\mathbf{p}(n)$, can be found from the initial probability vector by:

$$\mathbf{p}(n) = \mathbf{p}(0) \cdot \mathbf{P}(1) \cdot \ldots \cdot \mathbf{P}(n), \tag{1.25}$$

which can be seen by applying Equation 1.23 repeatedly. If the process is steady so that P is independent of n, this reduces to:

$$\mathbf{p}(n) = \mathbf{p}(0) \cdot \mathbf{P}^n. \tag{1.26}$$

1.2.2 Random Walk and the Diffusion Coefficient

We wish to compare a model for the dynamics of a particle based on the stochastic process presented in the previous section to Equation (1.10). We first consider the case where there is *no flow* through the tube. In that case the transport of a marked component will be solely by random molecular motion as discussed previously in this chapter.

If the velocity is zero, *i.e.* the first term on the right-hand-side of Equation (1.10) is zero, the equation reduces to what is known as "Fick's second law of diffusion":

$$\frac{\partial C(x,t)}{\partial t} = D\frac{\partial^2 C(x,t)}{\partial x^2}. \tag{1.27}$$

If the initial concentration is a Dirac Delta function in the y-z-plane at the origin $x = 0$ (i.e., they start at $x = 0$), the solution to Fick's second law of diffusion is:

$$C(x,t) = \frac{1}{\sqrt{\pi D t}} \exp\left(-\frac{x^2}{4Dt}\right). \tag{1.28}$$

We will show that Equation (1.27) is consistent with a stochastic model of random molecular motion (see also Lin and Segel [84] or a textbook in physical chemistry).

Consider the particle moving between the cells in discretized space in the x-direction as shown in Figure 1.9. Assign the following values to the transfer probabilities.

$$p_{ij} := \begin{cases} \frac{1}{2} & \text{for } j = i+1, \\ \frac{1}{2} & \text{for } j = i-1, \\ 0 & \text{otherwise.} \end{cases} \tag{1.29}$$

Thus the particle moves one cell at a time, with equal probability to the right and the left. Call the cell width Δ and the timestep ε.

One way to compare this stochastic model with Fick's second law is to consider its *continuous limit* in the following way. The transfer probabilities above are consistent with the following difference equation:

$$p(n+1,i) = \frac{1}{2}p(n,i-1) + \frac{1}{2}p(n,i+1), \tag{1.30}$$

remembering that $p(n,i)$ means the probability of the particle residing at $i\Delta$ at time $n\varepsilon$. Subtracting $p(n,i)$ from both sides and dividing through by $\Delta^2\varepsilon$ gives:

$$\frac{p(n+1,i) - p(n,i)}{\Delta^2\varepsilon} = \frac{p(n,i-1) + p(n,i+1) - 2p(n,i)}{2\Delta^2\varepsilon}. \tag{1.31}$$

Dividing through once more by Δ gives:

$$\frac{[p(n+1,i) - p(n,i)]/\Delta}{\Delta^2\varepsilon} = \frac{[p(n,i-1) + p(n,i+1) - 2p(n,i)]/\Delta}{2\Delta^2\varepsilon}. \tag{1.32}$$

Now allow Δ, ε to approach zero such that $\Delta^2/(2\varepsilon)$ remains constant and equal to D. Since the particle can take up any position, the probabilities divided by Δ then approach the *probability density*, $p(x,t)$, and the finite differences approach differentials giving:

$$\frac{\partial p(x,t)}{\partial t} = D\frac{\partial^2 p(x,t)}{\partial x^2}, \tag{1.33}$$

which we recognize as Fick's second law, Equation (1.27).

Another way is to compare the *solutions* of the stochastic model with that of the diffusion equation (1.27). We seek the probability distribution for the particle's position at a given time n. If the particle has taken r steps to the right its position is $2r - n$, and the probability of this is given by the binomial distribution.

$$P(X_n = 2r - n) = \frac{n!}{r!(n-r)!}\left(\frac{1}{2}\right)^n. \tag{1.34}$$

Introducing $i := (2r - n)$, so that $r = 1/2(i+n)$ and $(n-r) = 1/2(n-i)$, Equation(1.34) becomes:

$$P(X_n = i) = \frac{n!}{\left[\frac{1}{2}(n+i)\right]!\left[\frac{1}{2}(n-i)\right]!}\left(\frac{1}{2}\right)^n, \tag{1.35}$$

Taking natural logarithms of this and using Stirling's approximation: $\ln n! = (n+1/2)\ln n - n + \ln 2\pi^{1/2}$ gives:

$$\ln P(X_n = i) = \frac{1}{2}\Big[(\ln(\frac{2}{\pi n}) - (n+i+1)\ln(1+\frac{i}{n}) - \\ (n-i+1)\ln(1-\frac{i}{n})\Big]. \tag{1.36}$$

If $i/n \ll 1$, then $\ln(1 + i/n) \approx i/n$. Using this, simplifying and taking exponentials finally yields:

$$P(X_n = i) = \sqrt{\frac{2}{\pi n}} \exp\left(-\frac{i^2}{2n}\right). \tag{1.37}$$

This result was first obtained by Einstein in one of his papers from 1905 [42].

The distance travelled is $x = i\Delta$, and the time elapsed $t = n\varepsilon$. Substituting for i and n gives:

$$P(x) = \sqrt{\frac{2\varepsilon}{\pi t}} \exp\left(-\frac{x^2\varepsilon}{2\Delta^2 t}\right). \tag{1.38}$$

Equations (1.28) and (1.38) have the same form except for the trivial constant. Comparing the exponents leads to a relation between the diffusivity and the time and spatial step sizes:

$$D = \frac{\Delta^2}{2\varepsilon}. \tag{1.39}$$

We used this for relating the time step to the cell width when deriving the continuous limit, Equation (1.32) from Equation (1.33).

Figure 1.10 shows a comparison between a one-dimensional random walk process according to Equation (1.29) and the equivalent solution to Fick's second law of diffusion, Equation (1.28).

1.2.3 The One-Dimensional Convection-Diffusion Equation

To bring about a convective velocity in our stochastic model in addition to the diffusion, we can introduce a biased random walk model by allowing the transition probability to the right to be higher than that to the left. We can also allow the probability of remaining in the same cell non-zero, to give the probabilities of a classic *birth-death* chain:

$$p_{ij} := \begin{cases} \beta_i & \text{for } j = i+1, \\ \delta_i & \text{for } j = i-1, \\ \alpha_i & \text{for } j = i, \\ 0 & \text{otherwise.} \end{cases} \tag{1.40}$$

The reason for the name is that the chain represents a living system, a birth and a death corresponding to the transitions $p_{i,i+1}$, and $p_{i,i-1}$, respectively. The three transition probabilities sum, as mentioned, to 1: $\beta_i + \alpha_i + \delta_i = 1$.

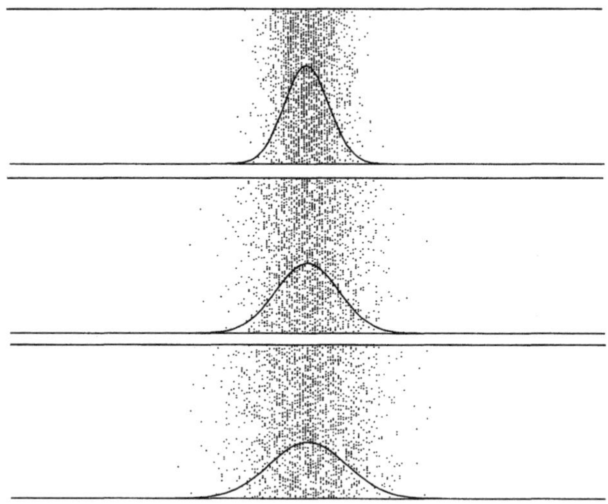

Figure 1.10: A one-dimensional random walk process for 300, 600 and 900 time-steps, respectively. The solid curves represent the solution to Ficks second law, *i.e.* Equation (1.28). If the parameters used are those for molecular movement in atmospheric air: $\Delta \approx 0.09\,\mu$m, $\varepsilon \approx 18$ ns, then the total time elapsed is 0.16 μs and tube sections shown are 40 μm long.

The full transition matrix for this chain is thus:

$$\begin{pmatrix} 1-\beta_1 & \beta_1 & 0 & 0 & 0 & \dots & 0 & 0 \\ \delta_2 & \alpha_2 & \beta_2 & 0 & 0 & \dots & 0 & 0 \\ 0 & \delta_3 & \alpha_3 & \beta_3 & 0 & \dots & 0 & 0 \\ \vdots & \vdots & \vdots & \vdots & \vdots & \ddots & \vdots & \vdots \\ 0 & 0 & 0 & 0 & 0 & \dots & \alpha_N & \beta_N \\ 0 & 0 & 0 & 0 & 0 & 0 & 0 & 1 \end{pmatrix}. \tag{1.41}$$

The first row represents a "reflecting boundary condition" and the last an "absorbing boundary condition". The boundary conditions will become more crucial in subsequent chapters.

Since we are here concerned with uniform pipe flow, we assume that the diffusivity, D and the velocity, v are independent of the position in the pipe and of time. We will later consider cases where this is not so.

In the previous section we assigned only probabilities of $\frac{1}{2}$ to upward and downward movements, and then introduced D by specifying the relative sizes of the space and time steps when considering the limit of $\Delta, \varepsilon \to 0$. In this case, however, it is more convenient to assign probabilities consistent with the physical process of convection/diffusion in advance. We assign the following probabilities to the movements in this process, which is of the "birth-death" type:

$$p_{ij} = \begin{cases} \beta_i = \frac{\varepsilon}{\Delta^2} D + \frac{\varepsilon}{2\Delta} v & \text{for } j = i + 1, \\ \alpha_i = 1 - \delta_i - \beta_i & \text{for } j = i, \\ \delta_i = \frac{\varepsilon}{\Delta^2} D - \frac{\varepsilon}{2\Delta} v & \text{for } j = i - 1, \\ 0 & \text{otherwise.} \end{cases} \tag{1.42}$$

With these choices of probability, the mean displacement per time step becomes εv and the mean square displacement $\Delta^2(\beta_i + \delta_i) = 2\varepsilon D$.

As in the previous section, we can write a difference equation based on the transfer probabilities:

$$p(n+1, i) = \beta_{i-1} p(n, i-1) + \alpha_i p(n, i) + \delta_{i+1} p(n, i+1). \tag{1.43}$$

Substituting from Equation (1.42):

$$p(n+1, i) = \left[\frac{\varepsilon}{\Delta^2} D + \frac{\varepsilon}{2\Delta} v\right] p(n, i-1) + \left[1 - 2\frac{\varepsilon}{\Delta^2} D\right] p(n, i) + \left[\frac{\varepsilon}{\Delta^2} D - \frac{\varepsilon}{2\Delta} v\right] p(n, i+1). \tag{1.44}$$

Again we can perform similar operations as in the previous section. Subtracting $p(n, i)$ from both sides of Equation (1.44), and dividing through by ε gives:

$$\frac{(p(n+1, i) - p(n, i)}{\varepsilon} = D \frac{p(n, i+1) - 2p(n, i) + p(n, i-1)}{\Delta^2} + v \frac{p(n, i+1) + p(n, i-1)}{\varepsilon}. \tag{1.45}$$

Dividing through again by Δ gives us, as in the previous section, the mean cell probability density from the absolute probabilities:

$$\frac{[p(n+1, i) - p(n, i)] / \Delta}{\varepsilon} = D \frac{[p(n, i+1) - 2p(n, i) + p(n, i-1)] / \Delta}{\Delta^2} + v \frac{[p(n, i+1) - p(n, i-1)] / \Delta}{\varepsilon}. \tag{1.46}$$

Allowing Δ, ε to approach zero causes Equation (1.46) to converge to:

$$\frac{\partial p(x,t)}{\partial t} = D\frac{\partial^2 p(x,t)}{\partial x^2} + v\frac{\partial p(x,t)}{\partial x}, \tag{1.47}$$

which we recognize as the one-dimensional convection-diffusion equation.

Chapter 2

Principles of Stochastic Process modeling

In this chapter we will introduce the basic technical concepts and tools of stochastic process modeling. We will concentrate on those processes that we have found useful in process technology, such as discrete Markov chains and diffusion processes. Our presentation will be application oriented, focussing on central ideas rather than on subtle mathematical details. In this spirit, we will generally delete the proofs or give only proofs of some special cases. Full proofs as well as technical details can be found in many textbooks on stochastic processes. Standard references are e.g. Karlin and Taylor ([72],[73]) and Bhattacharya and Waymire ([12]).

2.1 Stochastic Process Generalities

Stochastic processes are mathematical models for random phenomena evolving in time or space. Technically, a stochastic process is a family of random variables $(X_t)_{t \in I}$, indexed by the set I. In most of our models, the index set I has the interpretation of time. We have to distinguish continuous time models, where $I = [0, \infty)$ or $I = [0, T]$, from discrete time models, where $I = \mathbb{N}$ or $I = \{k\epsilon : k \in \mathbb{N}\}$, for some $\epsilon > 0$.

The random variables X_t take values in some space S, the so-called state space. The state space can be a subset of $\mathbb{R}^3$, e.g. in 3D models for the motion of particles in a chemical reactor, or a finite set, e.g. when the reactor is divided into finitely many compartments and X_t denotes the compartment occupied by the particle at time t. In general, X_t describes the state of the system at time t.

Technically, the random variables X_t are maps from an underlying sample space Ω into the state space S, i.e.

$$X_t : \Omega \to S.$$

The elements of Ω are called the outcomes. We can imagine that each $\omega \in \Omega$ contains full information about past, present and future development of the system under investigation. On the other hand, the specific ω that gives rise to the observations made by the experimental researcher is unknown and only revealed by the observations $X_t(w)$.

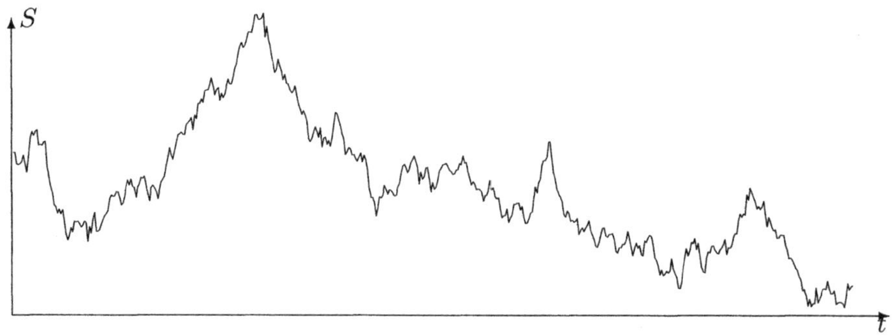

Figure 2.1: Sample path $(X_t(\omega))_{t\in\mathbf{I}}$ of a stochastic process with state space S and index set I.

For a given value of ω, the collection $(X_t(\omega))_{t\in I}$ is called a sample path of the process $(X_t)_{t\in I}$. One can think of a sample path as being one possible realization of the evolution of the system over time. Any sample path is an element of the space S^I, the space of all functions $f : I \to S$. Especially in continuous time models, we often have to impose some restrictions on the nature of the sample paths. One common restriction is to consider only continuous sample paths. The set of all possible paths allowed by the model is called path space.

The techniques for specification of a stochastic model vary significantly in the four possible combinations of discrete or continuous time and discrete or continuous state space. A convenient modeling strategy is to take a discrete time, discrete space model first and then take a limit by letting space and time discretization converge to zero. In this chapter we will introduce this modeling strategy by way of example in the section on approximation of diffusion processes.

As the stochastic process $(X_t)_{t\in I}$ has a different realisation for any $\omega \in \Omega$, we cannot provide a deterministic description of the path $(X_t(\omega))_{t\in I}$, e.g. by an equation of motion. For stochastic processes, the appropriate description is via the distribution of the process, assigning probabilities to all events that can be formulated in terms of the random variables $(X_t)_{t\in I}$. Stochastic process theory shows that it is sufficient to specify all finite dimensional distributions, i.e. to assign values to the probabilities

$$P(X_{t_1} \in A_1, \ldots, X_{t_n} \in A_n), \tag{2.1}$$

for all indices $t_1, \ldots, t_n \in I$ and all subsets $A_1, \ldots, A_n \subset S$. Probabilities of arbitrary events can then be computed by approximation with events depending only on a finite number of indices.

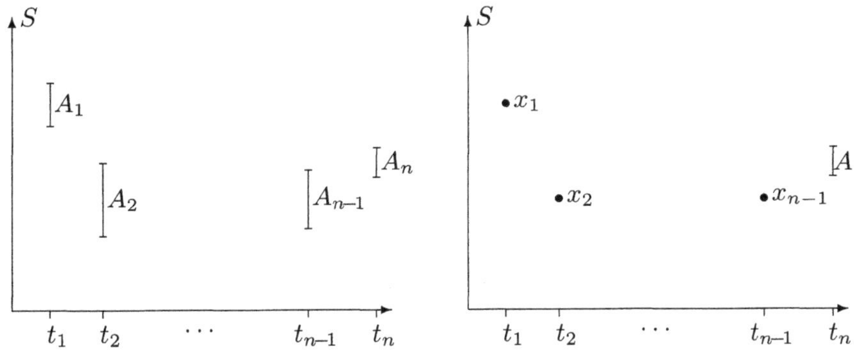

Figure 2.2: Specification of a stochastic process via n-dimensional marginal probabilities $P(X_{t_1} \in A_1, \ldots, X_{t_n} \in A_n)$ (left) and via transition probabilities $P(X_{t_n} \in A|X_{t_1} = x_1, \ldots, X_{n-1} = x_{n-1})$ (right).

The finite-dimensional distributions (2.1) can be calculated from the collection of all transition probabilities,

$$P(X_{t_n} \in A|X_{t_1} = x_1, \ldots, X_{t_{n-1}} = x_{n-1}), \tag{2.2}$$

for all indices $t_1, \ldots, t_n \in I$ with $0 \leq t_1 < \cdots < t_n$, all states $x_1, \ldots, x_{n-1} \in S$ and all sets $A \subset S$, together with the initial distribution, given by the probabilities $P(X_0 \in A)$, $A \subset S$. Transition probabilities have a clear physical interpretation. They specify the probability that the state of a process at time t_n lies in the subset A, given that it was equal to $x_1, \ldots, x_{n-1}$ at subsequent times $t_1, \ldots, t_{n-1}$ in the past.

Often we are just interested in the distribution of a process at fixed times $t \in I$, given by

$$P(X_t \in A),\ A \subset S.$$

These so-called one-dimensional marginal distributions, however, do not provide a full description of the distribution of the process, effectively because they do not give information about the way the states at different times are related.

In models with continuous state space, the one-dimensional marginal distributions often have a density function $p_t(x) = p(t,x)$, i.e. we can compute $P(X_t \in A)$ for any set $A \subset S$ by integration of $p(t,x)$ over A,

$$P(X_t \in A) = \int_A p(t,x)dx.$$

If the density $p(t,x)$ is continuous as a function of x, we get for a small neighborhood ΔV of x the approximation $P(X_t \in \Delta V) \approx |\Delta V|\, p(t,x)$, where $|\Delta V|$ denotes the volume of ΔV. Thus we have

$$p(t,x) \approx \frac{P(X_t \in \Delta V)}{|\Delta V|},$$

i.e. the density equals the probability that X_t lies in ΔV divided by the volume of ΔV.

When modeling particle transport in a reactor by a stochastic process, we look at the motion of a single particle. We thus refer to such models as microscopic models, in distinction with macroscopic models that refer to the behavior of a pulse of particles. In both types of models, the notion of particle density plays an important role, albeit with different interpretations. In a microscopic model, the particle density gives the probability to find a particle in a small volume element, divided by the volume. In a macroscopic model, particle density gives the fraction of particles found in a small volume element, again divided by the volume. The link between these two concepts is provided by the law of large numbers, which states that for a very large number of independently moving particles, the fraction of particles found in a given volume approximately equals the probability. Thus numerically, the two particle densities are equal, at least when the two assumptions are satisfied, namely that the number of particles in the pulse is very large and that they move independently.

2.2 Markov Processes

Markov processes are the simplest, nevertheless highly useful models for dependent observations. We say that the process $(X_t)_{t\in I}$ satisfies the Markov property, or that $(X_t)_{t\in I}$ is a Markov process, if the transition probabilities $P(X_{t_n} \in A | X_{t_1} = x_1, \ldots, X_{t_{n-1}} = x_{n-1})$ are a function of x_{n-1} only, and do not depend on $x_1, \ldots, x_{n-2}$, i.e. if

$$P(X_{t_n} \in A | X_{t_1} = x_1, \ldots, X_{t_{n-1}} = x_{n-1}) = P(X_{t_n} \in A | X_{t_{n-1}} = x_{n-1}) \tag{2.3}$$

holds for all time points $0 \le t_1 < \ldots < t_{n-1} < t_n$. This identity means that the future development of the process only depends on the present state and not on the past. In a colorful way, we can say that the process forgets its past. Whether this is a reasonable assumption, has to be judged by the scientist for any given experiment anew.

The distribution of a Markov process is completely specified by two pieces of information, namely

1. the distribution P_0 of X_0, i.e. the family of probabilities

   $$P_0(A) := P(X_0 \in A),\ A \subset S.$$

   P_0 is also called the initial distribution of the process.

2. the transition probabilities

   $$P_{s,t}(x, A) := P(X_t \in A | X_s = x),\ s, t \in I,\ x \in S,\ A \subset S.$$

   The function that maps the pair (x, A) to $P_{s,t}(x, A)$ is called the transition kernel of the process $(X_t)_{t \in I}$

We can then compute all finite-dimensional distributions of the process $(X_t)_{t \in I}$ by integration. For indices $t_1, \ldots, t_n \in I$ satisfying $t_1 < \ldots < t_n$ and all subsets $A_1, \ldots, A_n \subset S$ we get

$$\begin{aligned} &P(X_{t_1} \in A_1, \ldots, X_{t_n} \in A_n) \\ &= \int_S \int_{A_1} \cdots \int_{A_{n-1}} \int_{A_n} P_{t_{n-1},t_n}(x_{n-1}, dx_n) P_{t_{n-2},t_{n-1}}(x_{n-2}, dx_{n-1}) \ldots \\ &\qquad \ldots P_{0,t_1}(x_0, dx_1) P_0(dx_0). \end{aligned}$$

When viewed as a function of A while the argument x is held fixed, $A \mapsto P_{s,t}(x, A)$ defines a probability distribution on S, i.e. this set function is countably additive and satisfies $P_{s,t}(x, S) = 1$. As a function of x alone, $x \mapsto P_{s,t}(x, A)$ has to obey a mild technical condition, namely measurability, insuring that integrals $\int P_{s,t}(x, A) P(dx)$ can be meaningfully defined. These two properties together fully characterize a transition kernel.

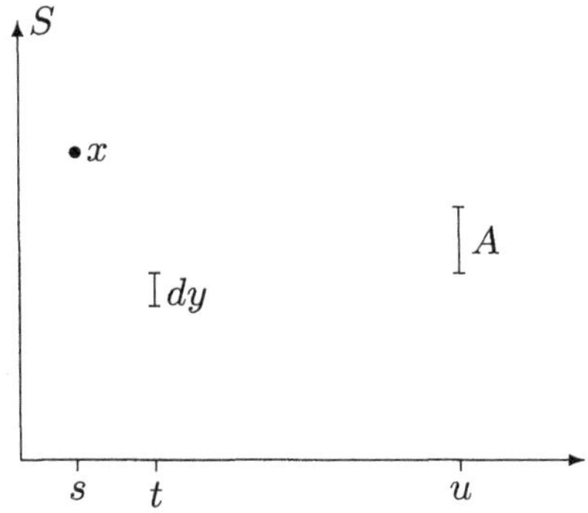

Figure 2.3: Chapman-Kolmogorov equations

The family of transition kernels satisfies the so-called Chapman-Kolmogorov equations that relate the transition kernels $P_{s,t}$, $P_{t,u}$ and $P_{s,u}$ for indices $s, t, u \in I$ satisfying $s < t < u$. We have

$$\begin{aligned} P_{s,u}(x, A) &= P(X_u \in A | X_s = x) \\ &= \int P(X_u \in A | X_t = y) P(X_t \in dy | X_s = x) \\ &= \int P_{t,u}(y, A) P_{s,t}(x, dy), \end{aligned}$$

as a result of the law of total probability.

In most physical processes, the transition laws are time homogeneous in the sense that the probability of a transition from x to A between times s and t only depends on the length of the time interval. In terms of the transition kernels this implies

$$P_{s+h,t+h}(x, A) = P_{s,t}(x, A),$$

for all $s, t, h \in I$, $s < t$. We then define

$$P_t(x, A) := P_{s,s+t}(x, A),$$

noting that the right-hand side does not depend on s. $P_t(x, A)$ gives the probability that, given that the present state of the system is x, it will be in A after t units of time have passed. The transition laws of a time homogeneous process are completely specified by the family of kernels $P_t(x, A)$, $t \geq 0$, as $P_{s,t}(x, A) = P_{t-s}(x, A)$. For a time homogeneous process, the

Chapman-Kolmogorov equation becomes

$$P_{s+t}(x, A) = \int P_t(y, A) P_s(x, dy)$$

Detailed analysis of Markov processes depends crucially on the character of the index set and of the state space. We will study Markov processes in discrete time and with discrete state space in the next section. The other type of Markov processes of interest to us are diffusion processes, in which case both index set and state space are continuous. Diffusion processes will be studied in a later section of this chapter.

2.3 Markov Chains

In this section we will concentrate on Markov processes in discrete time and with finite state space S. Such processes are called Markov chains. The elements of a discrete state space are commonly denoted by the letters h, i, j, k, and without loss of generality we may assume that $S = \{1, \ldots, N\}$. The index set is taken to be the set of all nonnegative integers $\mathbb{N} = \{0, 1, \ldots\}$, and we denote by X_n the state at time n. In discrete Markov models, the transition laws are completely specified by the transition probabilities

$$p_{m,n}(i, j) := P(X_n = j | X_m = i),$$

for $i, j \in S$ and $m, n \in \mathbb{N}$ satisfying $m \leq n$. By additivity of conditional probabilities, the transition probabilities $P_{m,n}(i, A)$ can then be computed by the identity $P_{m,n}(i, A) = P(X_n \in A | X_m = i) = \sum_{j \in A} p_{m,n}(i, j)$.

The Chapman-Kolmogorov equations for a Markov chain are

$$p_{l,n}(i, k) = \sum_{j \in S} p_{l,m}(i, j)\, p_{m,n}(j, k),$$

valid for all indices $l, m, n \in \mathbb{N}$ satisfying $l \leq m \leq n$ and all states $i, j, k \in S$. By iterated application of the Chapman-Kolmogorov equations we can compute all transition probabilities from the one-step transition probabilities $p_{n,n+1}(i, j) = P(X_{n+1} = j | X_n = i)$.

The Markov chain $(X_n)_{n \geq 1}$ is time homogeneous if and only if the one-step transition probabilities $p_{n,n+1}(i, j)$ do not depend on n. In most of our models this assumption will be made, and in this case we delete the index n and write

$$p_{ij} := P(X_{n+1} = j | X_n = i).$$

The matrix $P = (p_{ij})_{i,j \in S}$ is called the transition matrix of the Markov chain. Transition matrices are characterized by the properties that $p_{ij} \geq 0$

for all $i, j \in S$ and $\sum_{j \in S} p_{ij} = 1$ for all $i \in S$. Matrices satisfying these two properties are called stochastic matrices.

We define the m-step transition probabilities of a time homogeneous Markov chain by

$$p_{ij}^{(m)} := P(X_{n+m} = j | X_n = i),$$

again observing that the right hand side is independent of n. Moreover, we define the m-step transition matrix $P^{(m)} := (p_{ij}^{(m)})_{i,j \in S}$. From the Chapman-Kolmogorov equations we obtain the identity

$$p_{ik}^{(m)} = \sum_{j \in S} p_{ij}^{(m-1)} p_{jk}.$$

As a result we get the matrix identity $P^{(m)} = P^{(m-1)} P$, and by iterated application we find $P^{(m)} = P^m$, i.e. the m-step transition matrix equals the m-th power of the one-step transition matrix P.

The probability distribution of X_n can be computed from the transition matrix and the initial distribution by matrix multiplication. Suppose that $S = \{1, \ldots, N\}$, define the probability function of X_n,

$$p(n, i) := P(X_n = i), \quad 1 \leq i \leq N,$$

and the probability vector

$$p(n) := (p(n, 1), \ldots, p(n, N))$$

for any $n \in \mathbb{N}$. The we get by the law of total probability

$$\begin{aligned} p(n, j) = P(X_n = j) &= \sum_{i=1}^{N} P(X_{n-1} = i) \, P(X_n = j | X_{n-1} = i) \\ &= \sum_{j=1}^{N} p(n-1, i) \, p_{ij}, \end{aligned}$$

which can be expressed in vector-matrix notation as

$$p(n) = p(n-1) \, P. \tag{2.4}$$

Iterating this equation, we get the identity

$$p(n) = p(0) \, P^n.$$

We can thus compute the probability distribution at time n either directly by multiplying the initial probability distribution vector $p(0)$ from the right by P^n or recursively, by multiplying the time $(n-1)$ probability distribution vector $p(n-1)$ by P. Numerically, the computations involved in determining $p(n)$ can be carried out rather quickly using software such as MATLAB.

For many theoretical considerations, we are interested in analytic formulas for P^n and for $p(n)$, either exact or asymptotic. Exact formulas can be given if P is diagonalizable, i.e. if there exists an invertible matrix A such that

$$D := APA^{-1} \tag{2.5}$$

is a diagonal matrix with entries $\mu_1, \ldots, \mu_N$, say. From (2.5) we obtain the representation $P = A^{-1}DA$ and from here the following formula for the n-th power of the transition matrix,

$$P^n = (A^{-1}D\ A)(A^{-1}D\ A) \cdot \ldots \cdot (A^{-1}D\ A) = A^{-1}D^nA. \tag{2.6}$$

The right hand side is easily computed, since D^n is simply the diagonal matrix with entries $\mu_1^n, \ldots, \mu_N^n$ on the diagonal.

The transition matrix P can be diagonalized if and only if P has a full set of eigenvectors. Recall that $\lambda \in \mathbb{R}$ is called an eigenvalue of P if there exists a non-zero vector $u \in \mathbb{R}^K$ such that

$$u \cdot P = \lambda \cdot u.$$

In this case u is called a left eigenvector of P, associated with the eigenvalue λ. Note that for transition matrices it is natural to consider left eigenvectors, because this ties in with the recursion formula $p(n) = p(n-1)\,P$. We say that P has a full set of eigenvectors, if there exist K linearly independent eigenvectors of P. In that case P can be diagonalized as in (2.5) by the matrix with rows consisting of the linearly independent eigenvectors of P. The diagonal entries of D are the eigenvalues of P.

If P has a full set of eigenvectors, $u_1, \ldots, u_N$, say, we can write the initial distribution vector $p(0)$ as a linear combination of these vectors,

$$p(0) = \phi_1 u_1 + \ldots + \phi_N u_N \tag{2.7}$$

Then we get the following identity for the distribution vector of X_n,

$$\begin{aligned} p(n) &= p(0) \cdot P^n \\ &= \phi_1 \mu_1^n u_1 + \ldots + \phi_N \mu_N^n u_n, \end{aligned}$$

which is another way to look at diagonalizability of the transition matrix.

Only in a few special cases, a spectral decomposition (2.5) can be explicitly obtained. But still one can make statements about the behaviour of $p(n)$ and P^n for large values of n. This behaviour depends strongly on certain characteristics of the Markov chain which we shall introduce now.

Definition 1. *(i) Let $i, j, \in S$ be two states. Then j is said to be accessible from i, if there exists a non-negative integer n such that*

$$p_{ij}^{(n)} > 0,$$

i.e. if it is possible to reach j from i in a finite number of steps. We will use the symbol $i \to j$ to express that j is accessible from i.

(ii) We say that the states i and j communicate with each other if $i \to j$ and $j \to i$ hold simultaneously. In this case, we write $i \leftrightarrow j$.

The relation $\leftrightarrow$ is an equivalence relation on the state space S, i.e. it has the property that $i \leftrightarrow i$ (reflexivity), that $i \leftrightarrow j$ if and only if $j \leftrightarrow i$ (symmetry) and that $i \leftrightarrow j$ and $j \leftrightarrow k$ imply $i \leftrightarrow k$ (transitivity). The first two of these properties are obvious. To show transitivity, suppose that $i \leftrightarrow j$ and $j \leftrightarrow k$. Thus, by definition, there exist nonnegative integers m, n such that $p_{ij}^{(m)} > 0$ and $p_{jk}^{(n)} > 0$. Thus we can conclude that

$$p_{ik}^{(m+n)} = \sum_{h \in S} p_{ih}^{(m)} p_{hk}^{(n)} \geq p_{ij}^{(m)} p_{jk}^{(n)} > 0,$$

i.e. that k can be reached from i in $m + n$ steps, and hence $i \to k$. In the same way we can show that $k \to i$ and thus $i \leftrightarrow k$.

The state space S of a Markov chain thus splits into disjoint equivalence classes $S_1, S_2, \ldots$ in such a way that any two states from the same equivalence class communicate with each other and that states from two different equivalence classes do not communicate.

Definition 2. *A state $i \in S$ is called an absorbing state, if*

$$p_{ii} = 1,$$

and $p_{ij} = 0$ for all $j \in S$, $j \neq i$.

When modeling particle transport processes in process technology, we encounter essentially two kinds of Markov chains, which we have to analyse separately. In batch operated reactors, all states communicate with each other, i.e. there is just a single class of states. The particle distribution in such reactors will typically converge to a steady state. In continuously operated reactors, the exit state plays a special role as it does not communicate with the interior states of the reactor. All particles eventually leave the reactor. Questions of interest are also very different for the two types of reactors. In batch reactors one can try to compute the invariant distribution and the speed of convergence toward the invariant distribution. In continuously operated reactors, the time until a particle leaves the reactor is of interest.

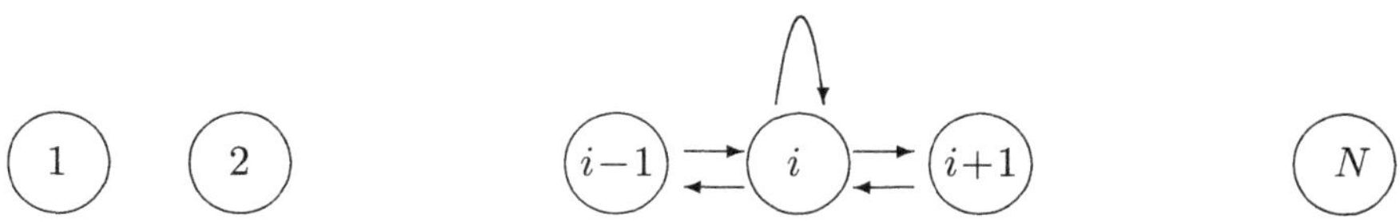

Figure 2.4: Possible transitions to and from state i in a birth-death Markov chain with state space $S = \{1, \ldots, N\}$.

Examples. (i) (*Birth-death process*) The state space of a birth-death process is either the set of non-negative integers or some finite interval of integers. For our purposes, we will take $S = \{1, \ldots, N\}$. A birth-death Markov chain is characterized by the fact that only three possible one-step transitions are allowed, namely remaining in the same state, moving one state ahead or one step back. If i denotes the present state of the process, we denote the probabilities for these three possible transitions by α_i, β_i and δ_i, respectively, where we assume that $\alpha_i + \beta_i + \delta_i = 1$. At the boundaries we require $\delta_1 = \beta_N = 0$. In this way we obtain the transition probabilities

$$p_{ij} = \begin{cases} \delta_i & j = i-1, i \neq 0 \\ \alpha_i & j = i \\ \beta_i & j = i+1, i \neq N \\ 0 & \text{otherwise} \end{cases}$$

and the resulting transition matrix

$$P = \begin{pmatrix} \alpha_1 & \beta_1 & 0 & 0 & \cdots & 0 & 0 & 0 & 0 \\ \delta_2 & \alpha_2 & \beta_2 & 0 & \cdots & 0 & 0 & 0 & 0 \\ 0 & \delta_3 & \alpha_3 & \beta_3 & \cdots & 0 & 0 & 0 & 0 \\ 0 & 0 & \delta_4 & \alpha_4 & \cdots & 0 & 0 & 0 & 0 \\ \vdots & \vdots & \vdots & \vdots & \ddots & \vdots & \vdots & \vdots & \vdots \\ 0 & 0 & 0 & 0 & \cdots & \alpha_{N-3} & \beta_{N-3} & 0 & 0 \\ 0 & 0 & 0 & 0 & \cdots & \delta_{N-2} & \alpha_{N-2} & \beta_{N-2} & 0 \\ 0 & 0 & 0 & 0 & \cdots & 0 & \delta_{N-1} & \alpha_{N-1} & \beta_{N-1} \\ 0 & 0 & 0 & 0 & \cdots & 0 & 0 & \delta_N & \alpha_N \end{pmatrix}.$$

Originally, birth-death chains were invented as models for the evolution of the size of a population. In the period between two transitions, we can either have one death, one birth or no change at all; the latter can also be the net result of one birth and one death occurring simultaneously. In particle transport applications, birth-death chains are a flexible class of discrete models for plug-flow with dispersion.

When modeling continuously operating reactors which allow inflow and outflow of material, we have to add the exterior of the reactor as an extra state. We usually denote this exit state by E or, for notational convenience by $N+1$, so that the state space enlargens to

$$S = \{1, \ldots, N, N+1\}.$$

In most practical applications, a particle that has left the reactor cannot return. This aspect is modeled by taking

$$\alpha_{N+1} = 1$$

and $\delta_{N+1} = \beta_{N+1} = 0$. In this model the exit state is thus an absorbing state. The full transition matrix is thus given by

$$P = \begin{pmatrix} \alpha_1 & \beta_1 & 0 & 0 & \cdots & 0 & 0 & 0 & 0 \\ \delta_2 & \alpha_2 & \beta_2 & 0 & \cdots & 0 & 0 & 0 & 0 \\ 0 & \delta_3 & \alpha_3 & \beta_3 & \cdots & 0 & 0 & 0 & 0 \\ 0 & 0 & \delta_4 & \alpha_4 & \cdots & 0 & 0 & 0 & 0 \\ \vdots & \vdots & \vdots & \vdots & \ddots & \vdots & \vdots & \vdots & \vdots \\ 0 & 0 & 0 & 0 & \cdots & \alpha_{N-2} & \beta_{N-2} & 0 & 0 \\ 0 & 0 & 0 & 0 & \cdots & \delta_{N-1} & \alpha_{N-1} & \beta_{N-1} & 0 \\ 0 & 0 & 0 & 0 & \cdots & 0 & \delta_N & \alpha_N & \beta_N \\ 0 & 0 & 0 & 0 & \cdots & 0 & 0 & 0 & 1 \end{pmatrix}$$

(ii) (*Ehrenfest urn model*) A rectangular box contains N identical balls. The box is divided up into two parts, say left and right half. At every time instant one ball is chosen at random and moved to the other part of the box. We are interested in the distribution of the balls over the two parts of the box. As the total number of balls remains fixed, the state of the system at time n is fully described by X_n, the number of balls in the left part of the box. The number of balls in the right part of the box is then $N - X_n$. We now compute the transition probabilities. Given that $X_n = i$, we choose with probability $\frac{i}{N}$ a ball from the left part and move it to the right part, and with probability $\frac{N-i}{N}$ we choose a ball from the right part of the box and move it to the left part. We thus get

$$p_{ij} = \begin{cases} \frac{i}{N} & \text{if } j = i+1 \\ \frac{N-i}{N} & \text{if } j = i-1 \\ 0 & \text{otherwise.} \end{cases}$$

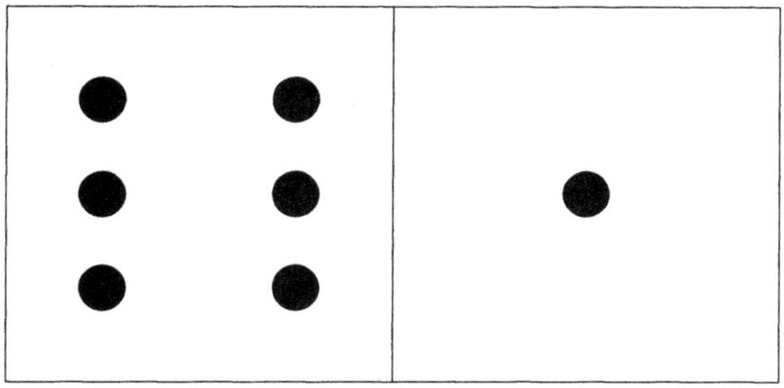

Figure 2.5: Ehrenfest urn model for the distribution of balls over the two parts a box; here for the case $N = 7$, $X_n = 6$.

At the boundaries we have $p_{12} = p_{N,N-1} = 1$. We thus get the transition matrix

$$P = \begin{pmatrix} 0 & 1 & 0 & \dots & 0 & 0 & 0 \\ \frac{1}{N} & 0 & \frac{N-1}{N} & \dots & 0 & 0 & 0 \\ 0 & \frac{2}{N} & 0 & \dots & 0 & 0 & 0 \\ \vdots & \vdots & \vdots & \ddots & \vdots & \vdots & \vdots \\ 0 & 0 & 0 & \dots & 0 & \frac{2}{N} & 0 \\ 0 & 0 & 0 & \dots & \frac{N-1}{N} & 0 & \frac{1}{N} \\ 0 & 0 & 0 & \dots & 0 & 1 & 0 \end{pmatrix}.$$

Observe that the Ehrenfest urn model is a special example of a birth-death process on the state space $\{0, \dots, N\}$, with $\alpha_i = 0$, $\beta_i = 1 - \frac{i}{N}$ and $\delta_i = \frac{i}{N}$.

The Ehrenfest urn model is a model for the diffusion of particles in a container, that was originally invented in order to understand the apparent contradiction between the entropy law of thermodynamics and the reversibility of diffusion processes. If the total number of particles in the container is large, we find eventually roughly half of the particles in one part and the other half in the other part of the container. In principle, a state where all particles are in one half can be reached, but it takes an incredibly long time for this to happen.

2.4 Long-Term Behavior of Markov Chains

In this section we consider a Markov chain $(X_n)_{n \in \mathbb{N}}$ in discrete time and with finite state space S. As in the previous sections, the transition probabilities are denoted by $p_{ij} = P(X_{n+1} = j | X_n = i)$ and the transition matrix by $P = (p_{ij})_{i,j \in S}$. We are most interested in the longterm behavior of the distribution of X_n, e.g. in possible convergence of $P(X_n = i)$ as $n \to \infty$. Suppose for a moment that

$$\pi_i = \lim_{n \to \infty} P(X_n = i)$$

exists for all $i \in S$, and consider the vector $\pi = (\pi_i)_{i \in S}$. Then we obtain

$$\begin{aligned}
\pi_j &= \lim_{n \to \infty} P(X_{n+1} = j) \\
&= \lim_{n \to \infty} \sum_{i \in S} P(X_n = i)\, p_{ij} \\
&= \sum_{i \in S} (\lim_{n \to \infty} P(X_n = i))\, p_{ij} = \sum_{i \in S} \pi_i\, p_{ij}.
\end{aligned}$$

Written in vector-matrix notation, this yields the following fundamental identity

$$\pi = \pi\, P. \tag{2.8}$$

Solutions of this equation play an important role in Markov chain theory, even in cases when $\lim_{n \to \infty} P(X_n = i)$ does not exist.

Definition 3. *A probability distribution $\pi = (\pi_i)_{i \in S}$ on the state space S is called a invariant distribution, if (2.8) is satisfied.*

The above considerations show that $\pi = (\lim_{n \to \infty} P(X_n = i))_{i \in S}$ defines an invariant distribution on S, provided the limit exists. Now, in the opposite direction, suppose that π is an invariant distribution. We can then take the Markov chain $(X_n)_{n \in \mathbb{N}}$ with initial distribution π and with transition matrix P. Then, according to (2.4), the distribution of X_n is given by

$$p(n) = \pi\, P^n = \pi,$$

and thus limit $\lim_{n \to \infty} P(X_n = i)$ trivially exists and equals π_i.

At this point, several questions arise naturally:

1. Under which conditions does a stationary distribution exist and when is it unique?

2. Under which conditions does $\lim_{n \to \infty} P(X_n = i)$ exist, and when is this limit independent of the initial distribution $p(0)$?

Examples. (i) The Ehrenfest urn model has a unique invariant distribution $\pi = (\pi_k)_{0\le k\le N}$, given by $\pi_k = \binom{N}{k}\frac{1}{2^N}$. Note that π is a binomial distribution on $\{0,\ldots,N\}$ with parameter $p = \frac{1}{2}$. That π is indeed an invariant distribution can be seen from

$$
\begin{aligned}
(\pi P)_k = \sum_{j=0}^{N} \pi_j p_{jk} &= \pi_{k-1}\, p_{k-1,k} + \pi_k\, p_{k+1,k} \\
&= \left(\binom{N}{k-1}\frac{N-(k-1)}{N} + \binom{N}{k+1}\frac{k+1}{N}\right)\frac{1}{2^N} \\
&= \left(\binom{N-1}{k-1} + \binom{N-1}{k}\right)\frac{1}{2^N} = \binom{N}{k}\frac{1}{2^N},
\end{aligned}
$$

where the last steps follow from some algebraic manipulations involving binomial coefficients, including the identity $\binom{N-1}{k-1} + \binom{N-1}{k} = \binom{N}{k}$, known from the Pascal triangle.

(ii) The birth death process with absorbing boundary at state $E = N+1$ has the invariant distribution $\pi = (0,\ldots,0,1)$. If $\beta_i \neq 0$ for all i, this is also the unique invariant distribution.

(iii) Consider a Markov chain with state space $S = \{1,2,3\}$ and transition probabilities $p_{12} = p_{21} = p_{33} = 1$ and all other p_{ij} equal to zero. The transition matrix is thus

$$
P = \begin{pmatrix} 0 & 1 & 0 \\ 1 & 0 & 0 \\ 0 & 0 & 1 \end{pmatrix}.
$$

This Markov chain has many invariant distributions, e.g.

$$
\pi_1 = (\frac{1}{2}, \frac{1}{2}, 0), \quad \pi_2 = (0,0,1),
$$

but also all convex combinations $\alpha\pi_1 + (1-\alpha)\pi_2$, $0 \le \alpha \le 1$. This Markov chain is not ergodic, in the sense of the following definition, because, e.g. the states 1 and 3 do not communicate.

Definition 4. *A Markov chain is called ergodic, if any two states $i,j \in S$ communicate with each other.*

In the literature on Markov chains, ergodic chains are also called irreducible. In the terminology introduced in Section 2.2, ergodic Markov chains have a single equivalence class of states.

Looking at the identity (2.8) from an algebraic point of view, an invariant distribution π is a left eigenvector of the transition matrix P with eigenvalue 1, with the additional properties that $\pi_i \geq 0$ and $\sum_{i\in S} \pi_i = 1$. Crucial is the non-negativity condition; the second condition can always be achieved by normalization. That 1 is an eigenvalue of P can, by the way, be directly seen from observing that $x = (1, \ldots, 1)^t$ is a right eigenvector with eigenvalue 1. Indeed, since P is a stochastic matrix, its row sums are equal to one and thus

$$P \begin{pmatrix} 1 \\ \vdots \\ 1 \end{pmatrix} = (\sum_{i\in S} p_{ij})_{j\in S} = \begin{pmatrix} 1 \\ \vdots \\ 1 \end{pmatrix}$$

Since the set of left eigenvalues coincides with the set of right eigenvalues, we may conclude that a left eigenvector with eigenvalue 1 exists. However, additional arguments are necessary if one wants to show that a left eigenvector with nonnegative entries exists. The algebraic theory of transition matrices is contained in the celebrated Perron-Frobenius theory of non-negative matrices.

Perron-Frobenius Theorem *Let $P = (p_{ij})_{1\leq i,j\leq N}$ be the transition matrix of an ergodic Markov chain.*

(i) The multiplicity of the eigenvalue 1 is one. There exists a unique left eigenvector $\pi = (\pi_i)_{1\leq i\leq N}$ with $\pi_i \geq 0$ and $\sum_{i=1}^N \pi_i = 1$.

(ii) All eigenvalues of P have modulus less than or equal to 1.

(iii) There exists a unique integer d such that the set of eigenvalues of P of modulus 1 equals the set of complex roots of unity

$$\lambda_k = e^{2\pi i \frac{k}{d}}, \quad 0 \leq k \leq d-1.$$

These eigenvalues are all simple.

Ergodicity is thus the key condition in connection with the first of the two questions raised above. Ergodic Markov chains have a unique invariant distribution. However, even for ergodic Markov chains, $\lim_{n\to\infty} P(X_n = i)$ need not exist for all initial distributions $p(0)$. However, a weaker form of convergence does always hold, namely convergence of Cesaro-averages, and the limit equals the invariant distribution. I.e. the right hand side of

$$\pi_i = \lim_{n\to\infty} \frac{1}{n} \sum_{k=0}^{n-1} P(X_k = i)$$

exists and defines a invariant distribution $\pi = (\pi_i)_{i\geq 1}$.

Example. Consider the Markov chain with state space $S = \{1, 2\}$ and transition probabilities $p_{12} = p_{21} = 1$ and $p_{11} = p_{22} = 0$. Thus the transition matrix is

$$P = \begin{pmatrix} 0 & 1 \\ 1 & 0 \end{pmatrix}.$$

Observe that P^n equals P or the identity matrix, according to whether n is odd or even. The Markov chain has a unique invariant distribution, namely $\pi = (\frac{1}{2}, \frac{1}{2})$. However, $\lim_{n\to\infty} P(X_n = 1)$ does not exist for all initial distributions $p(0)$. E.g. for $p(0) = (1, 0)$ we have

$$P(X_n = 1) = \begin{cases} 1 & \text{if } n \text{ is even} \\ 0 & \text{if } n \text{ is odd} \end{cases}$$

Existence of the limit $\lim_{n\to\infty} P(X_k = i)$ requires an additional condition which will be introduced now.

Definition 5. *Let $(X_n)_{n\in\mathbb{N}}$ be a Markov chain and $i \in S$. Then we define the period of i, denoted by d_i, as the greatest common divisor of the following set of integers*

$$R_i := \{n \leq 1 : p_{ii}^{(n)} > 0\}.$$

If $R_i = \emptyset$, we define $d_i = \infty$. The Markov chain is called aperiodic, if $d_i = 1$ for all $i \in S$.

One can show that communicating states have identical period. Thus, for an ergodic Markov chain all states have the same period, d say, which is also called the period of the Markov chain. The period of an ergodic Markov chain is exactly the integer d in part (iii) of the Perron-Frobenius theorem.

Looking back at the above example, we see that the Markov chain has period 2. A transition from any of the states back to itself is only possible in an even number of steps, and as a result $\lim_{n\to\infty} P(X_n = k)$ does generally not exist (it does exist, if we take the invariant distribution as initial distribution).

Ergodic theorem for aperiodic Markov chains. *Let $(X_n)_{n\geq 0}$ be an aperiodic ergodic Markov chain. Then there exists a unique invariant distribution $\pi = (\pi_i)_{i\in S}$ and*

$$\lim_{n\to\infty} P(X_n = i) = \pi_i,$$

for all $i \in S$. In fact, this convergence is geometrically fast, i.e.

$$\sum_{i \in S} |P(X_n = i) - \pi_i| \leq a\,\lambda^n$$

for some constants $a \in (0, \infty)$, $\lambda \in (0, 1)$.

Aperiodicity thus provides the key condition required to answer the second of the above two questions. For aperiodic ergodic Markov chains, $\lim_{n\to\infty} P(X_n = i)$ exists and is independent of the initial distribution $p(0)$. A key step in the proof of this result is the observation that for aperiodic ergodic Markov chains we can find an integer $m \geq 1$ such that

$$p_{ij}^{(m)} > 0 \quad \text{for all} \quad i, j \in S.$$

i.e. that the m-step transition matrix P^m has only positive entries.

For the study of non-ergodic Markov chains we need one extra piece of terminology.

Definition 6. *A state $i \in S$ is called essential, if for all $j \in S$ with $i \to j$ we also have $j \to i$ (and thus $i \leftrightarrow j$); i.e. if every state that can be reached from i communicates with i. Otherwise, the state is called inessential.*

If i and j are two communicating states, either both are essential or both are inessential. Thus, the states in one and the same equivalence class modulo the relation "$\leftrightarrow$" are either all essential or all inessential.

From an inessential state $i \in S$ one can get to a state j from where a return to i is impossible. As a result, one can prove that inessential states are visited only a finite number of times. Thus for inessential states, we have

$$\lim_{n\to\infty} P(X_n = i) = 0.$$

If the Markov chain $(X_n)_{n\geq 0}$ has only one equivalence class of essential states and if this class is aperiodic, then

$$\pi_i = \lim_{n\to\infty} P(X_n = i)$$

exists for all $i \in S$, and $\pi = (\pi_i)_{i\in S}$ is the unique invariant distribution. Moreover $\pi_i = 0$ for all inessential states $i \in S$.

Example. For the birth-death Markov chain with absorbing boundary at $N+1$, the states $1, \ldots, N$ are inessential states, if $\beta_i \neq 0$ for all i. Indeed, in this case $i \to N+1$, but $N+1$ does not lead to i. Thus $\pi = (0, \ldots, 0, 1)$ is the only invariant distribution and $\lim_{n\to\infty} P(X_n = i) = \pi_i$, i.e. eventually the process ends up in the absorbing state $N+1$.

2.5 Diffusion processes

In this section we will consider Markov processes in continuous time with general state space S. In most of our examples S will be some bounded or unbounded interval in $\mathbb{R}$. A Markov chain with stationary transition probabilities can be specified by the transition kernel

$$P_t(x, A) := P(X_{s+t} \in A | X_s = x),$$

where stationarity of the transition probabilities assures that the right hand side does not depend on s. Whereas in discrete time models it suffices to specify the one-step transition probabilities, we have to specify the entire family $(P_t)_{t\geq 0}$ of transition kernels in continuous time models. This task is complicated by the fact that the transition kernels have to obey the Chapman-Kolmogorov equations,

$$P_{s+t}(x, A) = \int P_t(y, A) P_s(x, dy),$$

valid for any $s, t \geq 0$. As a result, one can specify the family of transition kernels explicitly only in a few very simple examples.

Example. Historically, the first diffusion process investigated was Brownian motion, also known as Wiener process. Brownian motion is a stochastic process $(W_t)_{t\geq 0}$ with state space $\mathbb{R}$, index set $[0, \infty)$ and the following properties

1. $W_0 = 0$, and $(W_t)_{t\geq 0}$ has continuous sample paths.

2. For $0 = t_0 < t_1 < \ldots < t_n$, the increments $W_{t_0}, W_{t_1} - W_{t_0}, \ldots, W_{t_n} - W_{t_{n-1}}$ are independent random variables.

3. For $0 < s < t$, the increment $W_t - W_s$ has a normal distribution with mean 0 and variance $\sigma^2(t - s)$.

In the case when $\sigma^2 = 1$, we call $(W_t)_{t\geq 0}$ a standard Brownian motion or standard Wiener process.

Brownian motion is named after the Scottish botanist Robert Brown who discovered in 1827 under a microscope the motion of pollen suspended in liquid. Though the liquid seemed to be at rest, the pollen moved completely erratically, a fact that puzzled scientists in the 19th century. It was only in 1905 that Albert Einstein ([42]) could give a satisfactory explanation for Brownian motion. Einstein showed that Brownian motion is the result of collisions of the pollen with molecules of liquid. Denoting by

W_t one of the coordinates of the location of the pollen at time t, Einstein could derive the distribution of $W_t - W_s$ and show that it satisfies the above conditions. Moreover he obtained a formula relating the variance parameter σ^2 to physical parameters. Later Einstein's formula enabled Perrin to determine Avogadro's constant experimentally. For their discoveries, both Einstein and Perrin were awarded Nobel prizes. Norbert Wiener ([137]) could finally in 1923 rigorously prove existence of Brownian motion as a bona fide stochastic process. The difficult part in the existence proof is to show that the process $(W_t)_{t\geq 0}$ can be defined in such a way that the sample paths are continuous.

The transition densities of Brownian motion can be computed explicitly. Note that $W_{s+t} = (W_{s+t} - W_s) + W_s$, and that $W_{s+t} - W_s$ and W_s are independent. Thus, the conditional distribution of W_{s+t} given $W_s = x$ is a normal distribution with mean x and variance $\sigma^2 t$. Hence the transition density is

$$p_t(x,y) = \frac{1}{\sqrt{2\pi\sigma^2 t}} e^{-\frac{(y-x)^2}{2\sigma^2 t}}.$$

Continuous time Markov chains are usually specified by infinitesimal characteristics that completely determine the process, effectively because of the strong condition imposed by the Chapman-Kolmogorov equation. A helpful analogue is provided by the exponential function $f(x) = \exp(a \cdot x)$, where a is some arbitrary constant. The function f is completely specified by the functional equation $f(x+y) = f(x) \cdot f(y)$, by $f(0) = 1$ and by the value of its derivative at 0.

We introduce the transition operators $T_t, t \geq 0$, acting on functions $f : S \mapsto \mathbb{R}$ by

$$\begin{aligned} T_t f(x) : &= E(f(X_{s+t}) | X_s = x) \\ &= \int f(y) P_t(x, dy). \end{aligned}$$

The domain of the operator T_t consists of all those functions for which the integral on the right-hand side is well-defined. All non-negative functions are in the domain of T_t, and so are all bounded functions.

An application of the Chapman-Kolmogorov equation yields the following functional equation,

$$\begin{aligned} T_{s+t} f(x) &= \int f(y) P_{s+t}(x, dy) \\ &= \int\int f(y) P_t(z, dy) P_s(x, dz) = T_s(T_t f)(x), \end{aligned}$$

for all $s, t \geq 0$. If we denote by $\circ$ the usual composition of operators, we thus obtain

$$T_{s+t} = T_s \circ T_t,$$

i.e. $(T_t)_{t\geq 0}$ is a semigroup of operators, acting on a set of real-valued functions on the state space S. The operators T_t have special properties, first of all, $T_0(f) = f$ and $T_t(1) = 1$, where 1 denotes the function that is constant equal to 1. Moreover we have

$$T_t f \geq 0 \qquad \text{if } f \geq 0.$$

Definition 7. *Let $(X_t)_{t\geq 0}$ be a Markov process with transition operators $(T_t)_{t\geq 0}$ and let $f : S \to \mathbb{R}$. Then we define*

$$Af := \lim_{t\downarrow 0} \frac{T_t f - f}{t},$$

whenever this limit exists. Note that $Af : S \to \mathbb{R}$ and that $f \mapsto Af$ is a linear operator on the class of functions $f : S \to \mathbb{R}$ for which Af is defined. This linear operator is called infinitesimal generator of the Markov process.

Since $T_0 f = f$ we can rewrite the definition of the infinitesimal generator as

$$Af(x) = \lim_{t\downarrow 0} \frac{T_t f(x) - T_0 f(x)}{t} = \frac{\partial}{\partial t} T_t f(x)_{|t=0}.$$

The partial derivative of $T_t f(x)$, with respect to t, at an arbitrary point $t > 0$ can be computed with the help of the semigroup property as follows

$$\begin{aligned} \lim_{h\downarrow 0} \frac{T_{t+h} f(x) - T_t f(x)}{h} &= \lim_{h\downarrow 0} \frac{T_h(T_t f)(x) - (T_t f)(x)}{h} \\ &= A(T_t f)(x). \end{aligned}$$

In this way we obtain

$$\frac{\partial}{\partial t} T_t f(x) = A(T_t f)(x),$$

the so-called Kolmogorov backward equation for the family of transition operators $(T_t)_{t\geq 0}$.

From now on, we assume that $S \subset \mathbb{R}$ and that the transition kernel is absolutely continuous with respect to Lebesgue measure, i.e. that we have a transition density $p_t(x, y)$ satisfying

$$P_t(x, A) = \int_A p_t(x, y) dy.$$

Note that the Chapman-Kolmogorov equation for densities becomes

$$p_{t+s}(x,y) = \int p_s(x,z)p_t(z,y)dz,$$

and that the transition operator can be expressed as

$$T_s f(x) = \int f(z)p_s(x,z)dz.$$

If we consider, for fixed y, the function $f(z) := p_t(z,y)$, we thus obtain

$$T_s f(x) = \int p_t(z,y)p_s(x,z)dz = p_{t+s}(x,y),$$

and hence

$$\frac{p_{t+s}(x,y) - p_t(x,y)}{s} = \frac{T_s f(x) - f(x)}{s},$$

still with f defined as above. If we now let $s \to 0$ on both sides, we get

$$\frac{\partial}{\partial t} p_t(x,y) = Af(x).$$

Thus we have obtained the following theorem.

Theorem 8. *Under some regularity conditions, the family of transition densities $p_t(x,y)$ satisfies the Kolmogorov backward equation*

$$\frac{\partial}{\partial t} p_t(x,y) = Ap_t(x,y),$$

where A acts on $p_t(x,y)$ as a function of x.

There is another operator associated with a family of transition densities $(p_t(x,y))_{t\geq 0}$, which we shall introduce now. Given a function $g : S \to \mathbb{R}$, we define $T_t^* g : S \to \mathbb{R}$ by

$$T_t^* g(y) := \int p_t(x,y)g(x)dx,$$

again if the integral on the right hand side is defined. The operator T_t^* is the dual (or adjoint) operator to T_t, when we consider T_t and T_t^* acting on

L^2-functions. Indeed, we get for $f, g \in L^2$

$$\begin{aligned} < f, T_t^* g > &= \int f(y) T_t^* g(y)\, dy \\ &= \int f(y) \int p_t(x,y) g(x)\, dx\, dy \\ &= \int g(x) \int f(y) p_t(x,y) dy\, dx \\ &= \int T_t f(x)\, g(x) dx =< T_t f, g > \end{aligned}$$

where $< \cdot, \cdot >$ denotes the standard inner product on L^2.

The deeper meaning of the operator T_t^* is revealed by the observation that, if X_0 has the density g, then X_t has the density $T_t^* g$. Thus T_t^* expresses the way in which densities evolve in the course of time.

Subtracting $< f, g >$ from both sides of $< f, T_t^* g >=< T_t f, g >$ and dividing by t, we obtain

$$< f, \frac{T_t^* g - g}{t} >=< \frac{T_t f - f}{t}, g >$$

and thus, as $t \to 0$,

$$< f, \lim_{t \searrow 0} \frac{T_t^* g - g}{t} >=< Af, g >=< f, A^* g > .$$

From this identity we can infer that

$$\frac{\partial}{\partial t} T_t^* g(y)_{|t=0} = A^* g(y),$$

i.e. that A^* is the infinitesimal generator of the semigroup $(T_t^*)_{t \geq 0}$. We can now proceed similarly as in the proof of the Kolmogorov backward equation. We consider, for some fixed x, the function $g(z) = p_t(x,z)$. Then we get

$$T_s^* g(y) = \int g(z) p_s(z,y) dz = \int p_t(x,z) p_s(z,y) dz = p_{s+t}(x,y),$$

and hence

$$\frac{p_{t+s}(x,y) - p_t(x,y)}{s} = \frac{T_s^* g(y) - g(y)}{s}.$$

Thus, as $s \to 0$, we get $\frac{\partial}{\partial t} p_t(x,y) = A^* g(y)$, and we have proved the first part of the following theorem.

Theorem 9. *Under some regularity conditions, the family of transition densities $p_t(x,y)$ satisfies the Kolmogorov forward equation*

$$\frac{\partial}{\partial t}p_t(x,y) = A^* p_t(x,y),$$

where A^* acts on $p_t(x,y)$ as a function of y.
The density of X_t, denoted by $p(t,x)$, satisfies the Fokker-Planck equation

$$\frac{\partial}{\partial t}p(t,x) = A^* p(t,x),$$

where A^* operates on p as a function of the spatial variable x.

The Fokker-Planck equation can be derived in the same way as the Kolmogorov forward equation, by considering the function $g(z) = p(t,z)$ and observing that $T_s^* g(y) = p(t+s,y)$. Note that the proofs of all three differential equations have been largely heuristic, leaving out many difficult technical questions, such as concerning the domains of the operators T_t, T_t^*, A, A^*, and the frequent interchanges of integrals and partial derivatives.

We now turn our attention to diffusion processes. Roughly speaking, these are Markov processes with state space $S \subset \mathbb{R}^n$ and continuous sample paths. We will actually only consider diffusion processes whose state space is some, possibly unbounded, interval $S \subset \mathbb{R}$. In many ways, diffusion processes generalize Brownian motion. Like Brownian motion, they have continuous sample paths. Recall that the increments $\Delta W_t = W_{t+\Delta t} - W_t$ of Brownian motion have the properties

$$\begin{aligned} E(\Delta W_t) &= 0 \\ E((\Delta W_t)^2) &= \sigma^2 \Delta t. \end{aligned}$$

Increments of general diffusions will have mean and variance that depend on the present location of the process.

Definition 10. *A Markov process $(X_t)_{t\geq 0}$ is called a diffusion process with drift function $v(x)$ and diffusion function $D(x)$, if $(X_t)_{t\geq 0}$ has continuous sample paths and if*

$$\begin{aligned} E(X_{t+h} - X_t | X_t = x) &= v(x)\,h + o(h) \\ E((X_{t+h} - X_t)^2 | X_t = x) &= 2\,D(x)\,h + o(h) \\ E((X_{t+h} - Xt)^3 | X_t = x) &= o(h), \end{aligned}$$

as $h \to 0$.

It turns out that a diffusion process is completely specified by its drift and diffusion function. Moreover, given two functions $v(x)$, $D(x)$ that satisfy some regularity conditions, one can always construct a diffusion process with drift $v(x)$ and diffusion $D(x)$. There are two approaches to diffusion processes, the analytic treatment initiated by Kolmogorov ([78]) and Feller ([45]) and the probabilistic treatment due to Itô ([69]). The analytic treatment identifies the infinitesimal generator of the Markov process and shows existence of a semigroup with this generator. The probabilistic treatment represents a diffusion process as solution of the stochastic differential equation

$$dX_t = v(X_t)\,dt + \sqrt{2\,D(X_t)}\,dW_t, \tag{2.9}$$

where $(W_t)_{t\geq 0}$ is a Brownian motion. The theory of stochastic differential equations assures existence of a solution to (2.9) and shows that the solution is a diffusion process with drift $v(x)$ and diffusion $D(x)$, again assuming regularity conditions for $v(x)$ and $D(x)$.

The stochastic differential equation (2.9) cannot be interpreted in the obvious way, i.e. as

$$\frac{dX_t}{dt} = v(X_t) + \sqrt{2\,D(X_t)}\frac{dW_t}{dt},$$

because $(W_t)_{t\geq 0}$ does not have differentiable sample paths. Even worse, the sample paths of Brownian motion have unbounded total variation. In order to make sense of (2.9), we first rewrite it as an integral equation

$$X_t - X_0 = \int_0^t v(X_s)\,ds + \int_0^t \sqrt{2\,D(X_s)}\,dW_s. \tag{2.10}$$

In the next step, the terms on the right hand side of (2.10) have to be defined properly. The first of these terms is an ordinary integral, defined pathwise for each $\omega \in \Omega$ in the usual way as Riemann or Lebesgue integral. The second integral, however, cannot be defined pathwise, i.e. for each fixed $\omega \in \Omega$, effectively because $(W_t)_{t\geq 0}$ has unbounded total variation. Itô showed how to define integrals of the type $\int_0^t f(X_s)\,dW_s$, via a subtle L_2-approximation by appropriately defined Riemann sums. Such integrals are called stochastic integrals or Itô integrals. A stochastic process $(X_t)_{t\geq 0}$ is then said to be a solution of the stochastic differential equation (2.9) if it satisfies (2.10) for all $t \geq 0$.

The stochastic differential equation (2.9) can be interpreted as a stochastic perturbation of the ordinary differential equation

$$\frac{d}{dt}x(t) = v(x(t)).$$

The Euler scheme for numerical solutions of ordinary differential equations suggests an approximation scheme for solutions of the stochastic differential equation (2.9). We choose a time discretization ϵ and then solve (2.9) with initial value x_0 by the recursion scheme

$$X_{(n+1)\epsilon} = v(X_{n\epsilon})\,\epsilon + \sqrt{D(X_{n\epsilon})}\,\sqrt{\epsilon}\,Z_{n+1},$$

starting from $X_0 = x_0$. The process $(Z_n)_{n\geq 1}$ is a sequence of independent $N(0,1)$-valued random variables. The $\sqrt{\epsilon}\cdot Z$-term is motivated by the fact that $W_{(n+1)\epsilon} - W_{n\epsilon}$ has an $N(0,\epsilon)$-distribution, like $\sqrt{\epsilon}\cdot Z$ does. It can be shown that the stochastic Euler method indeed gives approximate solutions to the stochastic differential equation (2.9); for this and for other, usually faster, numerical solvers see the book of Kloeden and Platen ([76]).

In what follows, we shall calculate the infinitesimal generator A as well as its adjoint operator A^* for a diffusion process $(X_t)_{t\geq 0}$. Let f be a bounded three times continuously differentiable function. Then we get by a Taylor expansion of f about x,

$$\begin{aligned} T_t f(x) &= E(f(X_t)|X_0 = x) \\ &= f(x) + f'(x)E((X_t - x)|X_0 = x) \\ &\quad + \frac{1}{2}f''(x)E((X_t - x)^2|X_0 = x) \\ &\quad + \frac{1}{6}E(f'''(\xi_t)(X_t - x)^3|X_0 = x) \\ &= f(x) + t\,v(x)\,f'(x) + t\,D(x)\,f''(x) + o(t), \end{aligned}$$

where ξ_t is an intermediate point. From this we obtain

$$\lim_{t\to 0}\frac{1}{t}(T_t f(x) - f(x)) = v(x)\,f'(x) + D(x)\,f''(x).$$

Thus the infinitesimal generator of a diffusion process is a partial differential operator A, given by

$$Af(x) = v(x)\,f'(x) + D(x)\,f''(x).$$

By integration by parts, one can show that the adjoint operator A^* is given by

$$A^*g(y) = -\frac{d}{dy}(v(y)g(y)) + \frac{d^2}{dy^2}(D(y)g(y)).$$

We can now apply Theorem 8 and Theorem 9 to obtain the Kolmogorov forward and backward equation for the transition density of a diffusion process,

$$\begin{aligned}\frac{\partial}{\partial t}p_t(x,y) &= v(x)\frac{\partial}{\partial x}p_t(x,y) + D(x)\frac{\partial^2}{\partial x^2}p_t(x,y)\\ \frac{\partial}{\partial t}p_t(x,y) &= -\frac{\partial}{\partial y}(v(y)p_t(x,y)) + \frac{\partial^2}{\partial y^2}(D(y)p_t(x,y)).\end{aligned}$$

Again by applying Theorem 9 we obtain the so-called Fokker-Planck equation for the density $p(t,x)$ of X_t,

$$\frac{\partial}{\partial t}p(t,x) = -\frac{\partial}{\partial x}(v(x)p(t,x)) + \frac{\partial^2}{\partial x^2}(D(x)p(t,x)).$$

At the end of this section we will show how a diffusion process can be approximated to an arbitrary degree of accuracy by a time and space rescaled birth-death process. Such approximations are important, both from a theoretical as well as from an applied point of view. E.g., we will use birth-death process approximations for simulation of diffusion processes. For modeling purposes, it is often much easier to formulate a birth-death model than to specify a diffusion process. We will return to this topic in Chapter 9, where we will study approximations of processes that have both a jump and a diffusion part.

Let $v, D : [0,1] \to \mathbb{R}$ be the drift and diffusion functions; we assume that both are twice continuously differentiable. We define $D_0 = \sup_x D(x)$. To a given spatial discretization $\Delta > 0$ we define

$$\epsilon = \frac{\Delta^2}{2D_0},$$

which will be our time discretization step.

We start now from the birth-death Markov chain $(\widetilde{X}_n^{\Delta})_{n\geq 0}$ with state space $S = \{1, 2, \dots, \left[\frac{1}{\Delta}\right]\}$ and transition probabilities

$$\begin{aligned}\beta_i &= \frac{\epsilon}{\Delta^2}D(i\Delta) + \frac{\epsilon}{2\,\Delta}v(i\,\Delta)\\ \delta_i &= \frac{\epsilon}{\Delta^2}D(i\Delta) - \frac{\epsilon}{2\,\Delta}v(i\Delta)\\ \alpha_i &= 1 - 2\frac{\epsilon}{\Delta^2}D(i\Delta).\end{aligned}$$

We then define the continuous time process $(X_t^{\Delta})_{t\geq 0}$ by

$$X_t^{\Delta} := \Delta\, X_{[t/\epsilon]}, \quad t \geq 0.$$

$(X_t^{\Delta})_{t\leq 0}$ is a pure jump process making jumps at times that are integer multiples of ϵ and taking values $i\,\Delta$, $0 \leq i \leq \frac{1}{\Delta}$. Given that the state of the system at time $t = n\,\epsilon$ is $x = i\,\Delta$, it changes during the time interval $(n\,\epsilon, (n+1)\,\epsilon)$ to the new state $X^{\Delta}_{(n+1)\epsilon}$ according to the transition rules

$$\begin{aligned} P(X^{\Delta}_{(n+1)\epsilon} = x + \Delta | X^{\Delta}_{n\epsilon} = x) &= \frac{\epsilon}{\Delta^2} D(i\Delta) + \frac{\epsilon}{2\Delta} v(i\Delta) \\ P(X^{\Delta}_{(n+1)\epsilon} = x - \Delta | X^{\Delta}_{n\epsilon} = x) &= \frac{\epsilon}{\Delta^2} D(i\Delta) - \frac{\epsilon}{2\Delta} v(i\,\Delta) \\ P(X^{\Delta}_{(n+1)\epsilon} = x | X^{\Delta}_{n\epsilon} = x) &= 1 - 2\frac{\epsilon}{\Delta^2} D(i\Delta). \end{aligned}$$

From these transition probabilities we obtain the following identities for the first two conditional moments of the increments of the process $(X^{\Delta}_t)_{t\geq 0}$,

$$\begin{aligned} E((X^{\Delta}_{(n+1)\epsilon} - X^{\Delta}_{n\epsilon}) \,|\, X^{\Delta}_{n\epsilon} = x) &= \epsilon\, v(x) \\ E((X^{\Delta}_{(n+1)\epsilon} - X^{\Delta}_{n\epsilon})^2 \,|\, X^{\Delta}_{n\epsilon} = x) &= 2\,\epsilon\, D(x). \end{aligned}$$

We see that these conditional moments coincide with those of a diffusion process with drift and diffusion function $v(x)$ and $D(x)$, respectively.

One can show that for $\Delta \to 0$ the sequence $(X^{\Delta}_t)_{t\geq 0}$ converges in distribution to a diffusion process $(X_t)_{t\geq 0}$ with drift and diffusion functions $v(x)$ and $D(x)$, respectively. The full proof is technically involved and will not be given here. We will consider one aspect, namely formal convergence of the Fokker-Planck equation of the discrete process $(X^{\Delta}_t)_{t\geq 0}$ to the Fokker-Planck equation of a diffusion process. Let

$$p_{\Delta}(n\,\epsilon, i\,\Delta) = \frac{1}{\Delta} P(X_{n\,\epsilon} = i\,\Delta),$$

i.e. $p_{\Delta}(n\epsilon, i\Delta)$ gives the approximate density at time $n\,\epsilon$ in location $i\,\Delta$. In order to shorten notation, we will drop the subscript Δ in what follows and simply write $p(n\,\epsilon, i\,\Delta)$. Then we have

$$\begin{aligned} &P(X_{(n+1)\epsilon} = i\Delta) \\ &\quad = \Big(\frac{\epsilon}{\Delta^2} D((i-1)\Delta) + \frac{\epsilon}{2\Delta} v((i-1)\Delta)\Big)\, P(X_{n\epsilon} = (i-1)\Delta) \\ &\quad\quad + \Big(1 - 2\frac{\epsilon}{\Delta^2} D(i\Delta)\Big)\, P(X_{n\epsilon} = i\Delta) \\ &\quad\quad + \Big(\frac{\epsilon}{\Delta^2} D((i+1)\Delta) - \frac{\epsilon}{2\Delta} v((i+1)\Delta)\Big)\, P(X_{n\epsilon} = (i+1)\Delta) \end{aligned}$$

and thus

$$\begin{aligned} &\frac{1}{\epsilon}(p((n+1)\,\epsilon, i\Delta) - p(n\,\epsilon, i\Delta)) \\ &\quad = \frac{1}{\Delta^2} D((i-1)\Delta) p(n\epsilon, (i-1)\Delta) - 2D(i\Delta) p(n\epsilon, i\Delta) \\ &\qquad\qquad\qquad + D((i+1)\Delta) p(n\epsilon, (i+1)\Delta)) \\ &\quad - \frac{1}{2\Delta}\,(v((i+1)\Delta) p(n\epsilon, (i+1)\Delta) - v(i-1)\Delta) p(n\epsilon, (i-1)\Delta)) \end{aligned}$$

If we let $\Delta \to 0$, and thus $\epsilon \to 0$, the left hand side converges to $\frac{\partial}{\partial t}p(t,x)$ and the right hand side to

$$-\frac{\partial}{\partial x}(v(x)\,p(t,x)) + \frac{\partial^2}{\partial x^2}(D(x)\,p(t,x)),$$

so that we obtain the partial differential equation

$$\frac{\partial}{\partial t}p(t,x) = -\frac{\partial}{\partial x}(v(x)\,p(t,x)) + \frac{\partial^2}{\partial x^2}(D(x)\,p(t,x))$$

which is the Fokker-Planck equation for the diffusion process with drift $v(x)$ and diffusion $D(x)$.

2.6 First Exit Times and RTD Curves

In this section we consider processes with absorbing boundaries. These will be either Markov chains $(X_n)_{n\in\mathbb{N}}$ with state space $S = \{1,\ldots,N,N+1\}$ and absorbing boundary at $N+1$ or diffusion processes $(X_t)_{t\geq 0}$ with state space $S = [0,1]$ and absorbing boundary at 1. We are interested in the time until the processes reach their absorbing state, in which they then remain forever. This time is a random variable, called first exit time and denoted by the letter T. Denoting the absorbing state generically by the letter E we thus get

$$T = \inf\{t \geq 0 : X_t = E\}.$$

We further define the first exit time distribution function by

$$F(t) = P(T \leq t).$$

In continuous models, T will often have a density $E(t)$ which we will call the first exit time density. We then have $F(t) = \int_0^t E(s)ds$ and $E(t) = \frac{d}{dt}F(t)$. In discrete time models, the analogue of the density function is the probability function $E(n) = P(T = n)$. We then have the relations $F(n) = \sum_{0\leq k\leq n} E(k)$ and $E(n) = F(n) - F(n-1)$.

First exit times play a very important role when we are modeling continuously operated reactors where particles enter the reactor, remain inside for some time, and eventually exit the reactor. In this context, the first exit time is called the residence time, its distribution is called the residence time distribution (RTD) and its distribution function the RTD curve. The residence time distribution is a very important characteristic of continuously operated reactors as particles typically undergo some reaction while

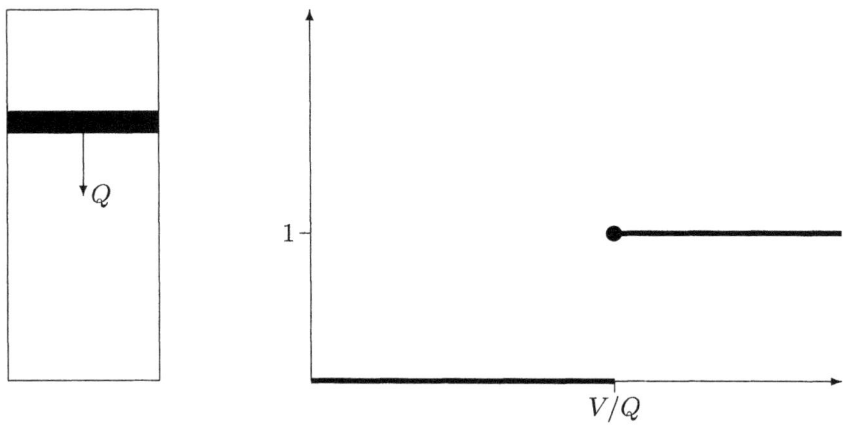

Figure 2.6: Plug-flow of a pulse of marked particles moving from top to bottom through the reactor (left) and corresponding RTD curve (right).

they are inside the reactor, and the intensity of this reaction depends crucially on the residence time. How close the residence time predicted by a mathematical model is to the actual, experimentally observed, residence time, is an important criterion for the quality of a model.

Above, we have presented the microscopic view of residence time distributions, focussing on the residence time T of a single particle and its distribution and density function. In the process technology literature, a macroscopic point of view is more common, focussing on the motion of a pulse of marked particles injected into the reactor at one time. The RTD curve $F(t)$ is then defined as the fraction of particles in the pulse that have left the reactor after t units of time. If the individual particles move independently, both approaches yield numerically the same RTD curve, by the law of large numbers.

In general, it is quite difficult to give exact analytic formulae for the first exit time distribution, and we have to rely on numerical computations or on asymptotic formulae. First of all we will give some examples where exact formulae can be obtained.

Example 1: Plug-flow. Consider a cylindrical reactor with unit cross sectional area of height, and thus volume, V. Particles enter the reactor at the top with an inflow rate Q and exit at the bottom at the same rate. Plug-flow is characterized by the fact that the particles move inside the reactor vertically from top to bottom at constant speed Q. In this way, a pulse of marked particles, entering the reactor at one time, moves like a plug through the reactor, see Figure 2.6.

If we denote by X_t the vertical location of a particle at time t, measured from the top of the reactor, we obtain

$$X_t = Q\,t, \quad t \geq 0.$$

Here we have assumed that the particle enters the reactor of time $t = 0$, i.e. that $X_0 = 0$. The residence time of plug-flow is then

$$T = \inf\{t : X_t = V\} \equiv \frac{V}{Q}.$$

As $(X_t)_{t\geq 0}$ is a deterministic process, the residence time is deterministic, too. The residence time distribution is a point mass at $\frac{V}{Q}$ with distribution function.

$$F(x) = \begin{cases} 0 & \text{if } x < V/Q \\ 1 & \text{if } x \geq V/Q. \end{cases}$$

Example 2: Ideal Mixer. In an ideal mixer, particles inside the reactor are instantaneously uniformly distributed over the entire reactor. We consider a reactor of volume V and volumetric inflow Q. For a discrete model of an ideal mixer, we subdivide the reactor into N horizontal cells of equal volume. The cells are numbered $1, \ldots, N$, and we take an extra cell with index $E = N + 1$ to denote the exit state. We model the location of a particle at integers multiples of the time unit and denote by X_n the index of the cell occupied by the particle after n transitions. We take $(X_n)_{n\geq 1}$ to be a Markov process with initial state $X_0 = 1$ and transition probabilities

$$p_{ij} = \begin{cases} \frac{1-Q/V}{N} & \text{if } 1 \leq j \leq N \\ Q/V & \text{if } j = N+1, \end{cases}$$

for $1 \leq i \leq N$ and $p_{N+1,N+1} = 1$. We can thus imagine that at any time n we toss a coin that shows Head with probability Q/V and Tail with probability $1 - Q/V$. If Head occurs, the particle is instantly removed from the reactor. This happens with probability Q/V, in line with the assumption that the volumetric inflow is v. If Tail occurs, the particle has equal chance to move to one of the K interior cells of the reactor, reflecting the ideal mixing assumption.

If we denote the sequence of coin tosses by $(\xi_i)_{i\geq 1}$, the residence time becomes

$$T = \inf\{n : \xi_n = \text{Head}\}.$$

The probability function of T is given by

$$\begin{aligned} P(T = k) &= P(\xi_1 = \ldots = \xi_{k-1} = \text{Tail}, \xi_k = \text{Head}) \\ &= \left(1 - \frac{Q}{V}\right)^{k-1} \frac{Q}{V}, \quad k \geq 1. \end{aligned}$$

Thus T has a geometric distribution with parameter $\frac{Q}{V}$. The mean and the variance of T are given by $E(T) = \frac{V}{Q}$ and $\text{Var}(T) = \frac{V^2}{Q^2}$, respectively.

In order to obtain a model in continuous time, we let the time step converge to zero and adapt the transition probabilities accordingly. If a transition occurs at integer multiples of Δt, we take the transition probabilities

$$p_{ij} = \begin{cases} \frac{1-Q\,\Delta t/V}{N} & \text{if } 1 \leq j \leq N \\ \frac{Q\,\Delta t}{V} & \text{if } j = N+1. \end{cases}$$

The residence time T now has the probability function

$$P(T = k\,\Delta t) = \left(1 - \frac{Q\Delta t}{V}\right)^{k-1} \frac{Q\Delta t}{V}.$$

Thus

$$P(t < T \leq t + \Delta t) \approx \left(1 - \frac{Q\,\Delta t}{V}\right)^{t/\Delta t} \frac{Q\Delta t}{V} \approx \frac{Q}{V} e^{-Qt/V}\,\Delta t.$$

In the limit, as $\Delta t \to 0$, we get the density $E(t) = \frac{Q}{V} e^{-Qt/V}$ for T. This is the density of an exponential distribution with parameter Q/V. The RTD curve is thus given by

$$F(t) = \int_0^t \frac{Q}{V} e^{-\frac{Q}{V}s} ds = 1 - e^{-\frac{Q}{V}t},$$

for $t \geq 0$. Observe that F is the unique solution to the differential equation

$$F'(t) = -\frac{Q}{V}\,(1 - F(t))$$

with boundary condition $F(0) = 0$. This differential equation can also be obtained from a macroscopic consideration of residence times in an ideal mixer. Indeed, consider a pulse of marked particles entering the reactor at time $t = 0$. At time t, the fraction $1 - F(t)$ of the marked particles is still in the reactor. Of these particles the fraction $\frac{Q}{V}\Delta t$ leaves in the time interval $[t, t + \Delta t]$. Thus $F(t + \Delta t) - F(t) = -(\frac{Q}{V}\Delta t)(1 - F(t))$, leading to the above differential equation.

Example 3: Sequence of ideal mixers. Suppose we are given n ideal mixers in a row, each having volume V/n. Particles enter the first mixer, stay for the length of one residence time T_1, move to the next mixer, stay for T_2, and so on until they finally exit from the last mixer. The total residence time is thus a sum $T = T_1 + \ldots + T_n$ of n independent residence times in the individual mixers. The distribution of T is the convolution of the distributions of $T_1, \ldots, T_n$. In the case of a discrete time sequence of ideal mixers we thus obtain a negative-binomially distributed residence time with probability function

$$P(T = k) = \binom{k-1}{n-1}\left(1 - \frac{nQ}{V}\right)^{k-n}\left(\frac{nQ}{V}\right)^n, \qquad k = n, n+1, \ldots$$

In the continuous case, we obtain a Gamma$(n, \frac{nQ}{V})$-distributed residence time with density

$$f_T(t) = \frac{(nQ/V)^n}{(n-1)!} t^{n-1} e^{-(nQ/V)t}\, 1_{[0,\infty]}(t).$$

Observe that in all three examples above, the mean residence time satisfies

$$E(T) = \frac{V}{Q}.$$

This identity, known as Danckwerts' law, hold universally for continuously operated reactors, as was shown by Danckwerts ([30]). We will show in Chapter 4 that stochastic transport models satisfy Danckwerts' law, provided some natural conditions are satisfied.

We will now derive an explicit formula for the residence time distribution in a finite state Markov transport model that can be used for numerical calculations. Let the state space be $S = \{1, 2, \ldots, N, N+1\}$, where again $N + 1$ denotes the exterior of the reactor. We consider a Markov chain $(X_n)_{n \geq 0}$ with transition matrix $P = (p_{ij})_{1 \leq i,j \leq N+1}$. If the initial distribution of the particle is given by $p(0) = (p(0,1), \ldots, p(0, N+1))$, the distribution after n transitions is given by

$$p(n) = p(n-1) \cdot P = \ldots = p(0) \cdot P^n.$$

We denote the i–th coordinate of $p(n)$ by $p(n, i)$, i. e.

$$p(n) = (p(n,1) \ldots, p(n, N+1)).$$

Since $N + 1$ is an absorbing boundary, a particle has a residence time less than or equal to n if and only if it is in state $N + 1$ at time n. As a consequence we obtain

$$P(T \leq n) = P(X_n = N + 1) = p(n, N + 1). \tag{2.11}$$

and

$$P(T = n) = p(n, N + 1) - p(n - 1, N + 1).$$

The residence time distribution function $F(n) := P(T \leq n)$ may thus be calculated numerically as the last coordinate of $p(n)$, the probability vector of the state at time n.

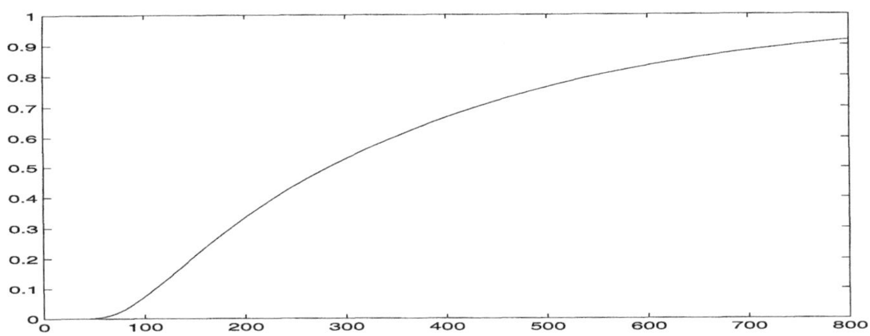

Figure 2.7: RTD curve for a particle transport model in a fluidized bed reactor, see Chapter 5 for details of the model.

The RTD curves in many transport models have a characteristic exponential tail, as is displayed e.g. in Fig. 2.7. This behavior is also found in many experimental data. In order to investigate this phenomenon, we consider a Markov chain with state space $\{1, \ldots, N, N+1\}$, where $E = N+1$ is an absorbing boundary and where the states $1, \ldots, N$ are inessential states. Moreover, we assume that the transition matrix P has a full set of left eigenvectors $u_1, \ldots, u_{N+1}$ We may assume that the corresponding eigenvalues $\mu_1, \ldots, \mu_{N+1}$ are arranged in decreasing order of magnitude, i.e. that

$$|\mu_1| \geq |\mu_2| \geq \ldots \geq |\mu_{N+1}|.$$

According to the Perron-Frobenius theorem, the largest eigenvalue of P is 1, and the associated eigenvector is the unique invariant distribution.

Hence $\mu_1 = 1$ and

$$u_1 = \pi = (0, \dots, 0, 1).$$

Moreover, since the Markov chain is aperiodic, all other eigenvalues have a modulus strictly less than one, i.e.

$$|\mu_i| < 1, \quad i = 2, \dots, N+1.$$

By assumption, the eigenvectors $u_1, u_2, \dots, u_{N+1}$ form a basis of $\mathbb{R}^{N+1}$. Thus we can express any initial distribution $p(0)$ as linear combination of the eigenvectors,

$$p(0) = a_1 u_1 + \dots + a_{N+1} u_{N+1},$$

for some $a_1, \dots, a_{N+1} \in \mathbb{R}$. Since $u_i\, P = \mu_i\, u_i$, we get the following identity

$$\begin{aligned} p(n) = p(0)\, P^n &= a_1 u_1\, P^n + a_2 u_2 P^n + \dots + a_{N+1} u_{N+1}\, P^n \\ &= a_1\, \pi + \mu_2^n a_2 u_2 + \dots + \mu_{N+1}^n\, a_{N+1} u_{N+1}. \quad (2.12) \end{aligned}$$

Letting $n \to \infty$ and using the fact that $|\mu_i| < 1$ for all $i \in \{2, \dots, N+1\}$, we get $\lim_{n\to\infty} p(n) = a_1\, \pi$. On the other hand, since π is the invariant distribution of the Markov chain, we have $\lim_{n\to\infty} p(n) = \pi$. Thus $a_0 = 1$, and we may rewrite (2.12) to get

$$p(n) = \pi + \mu_2^n\, a_2\, u_2 + \sum_{i=3}^{N+1} \mu_i^n\, a_i\, u_i.$$

Under the additional assumption that $|\mu_2| < |\mu_1|$, the first two terms on the right hand side dominate and thus

$$p(n) \approx \pi + \mu_2^n a_2 \mu_2. \quad (2.13)$$

Now, since $\pi = (0, \dots, 0, 1)$, we get from (2.13) the following expressions for the coordinates of $p(n)$,

$$\begin{aligned} p(n, i) &\approx \mu_2^n\, a_2\, u_2(i) \quad , \quad 1 \le i \le N \\ p(n, N+1) &\approx 1 + \mu_2^n\, a_2\, u_2(N+1) = 1 - a\, \mu_2^n, \end{aligned}$$

where we have put $a := -a_1 \mu_1(N)$. Here $u_2(i)$ denotes the i-th coordinate of u_2. Observe that $a \ge 0$, since $p(n, N+1) \le 1$. We thus get the following asymptotic expression for the RTD-curve

$$F(n) \approx 1 - a \mu_2^n.$$

This is the typical exponential decay of the RTD-curve for large n, found in experimental results in fluidized beds.

Chapter 3

Batch Fluidized Beds

In this chapter we look at a specific application of stochastic models for batch systems, namely that of vertical particle transport in batch gas fluidized beds. Fluidized beds are complex systems where processes, some of which are discrete in nature, take place on different scales. They are therefore notoriously difficult to model. We will show how the strategy of stochastic modeling can be used to model the transport processes, even for beds where the traditional conservation equation methods fail to produce a model. This chapter will also show how stochastic models are useful even for systems that, due to the many particles present, behave in an essentially deterministic way.

3.1 Flow Regimes

In gas fluidized beds a gas is injected through a porous or perforated plate, the "gas distributor" that supports a bed of particles (for instance sand or catalyst particles), see Figure 3.1. If the volumetric flow (or superficial gas velocity[1], U, as is most often used) of the gas is sufficient, the weight of the particles, less the Archimedian boyancy force acting on them, will be supported by the flow force of the gas. The particles are then said to be fluidized. Beyond the superficial gas velocity just required to fluidize the particles, which is called the *minimum fluidization velocity*, U_{mf}, the bed of particles exhibits a fluid-like character, where the particles are only loosely in contact.

When continuing to increase the superficial gas velocity, the bed may or may not go through a regime of "particulate expansion", where it expands uniformly. At a certain gas velocity, often called the *minimum bubbling*

[1]the gas velocity as it would be in an empty column

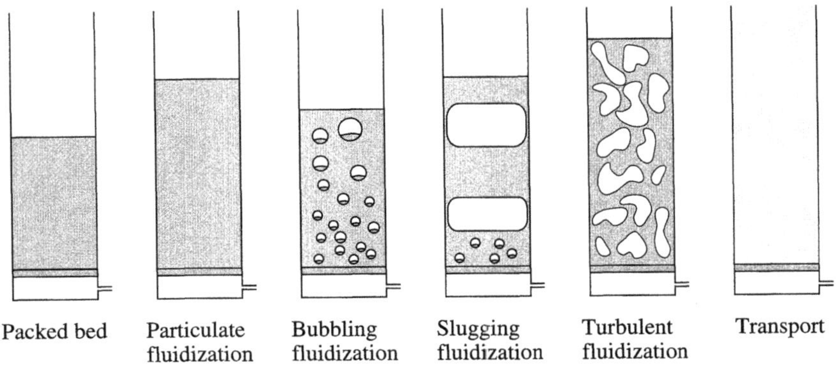

Figure 3.1: Fluidization regimes

velocity, U_{mb}, bubbles will begin to appear, much like vapor bubbles in a boiling liquid. As one moves up the bed, the total flow of gas in the empty part of the bubbles will remain almost constant but the individual fluidization bubbles will grow, due to coalescence, as suggested in the figure.

Further increase in the gas velocity will lead to the formation of large, slowly moving voids almost filling the bed cross-section, so-called "slugs", and eventually to a chaotic mixture of particle-dense and particle-void regions, known as "turbulent fluidization". Finally, if the gas velocity is high enough, the bed particles will be transported upward. If a reactor is operated in this latter regime it is often referred to as a "riser".

The particles in batch fluidized beds, as we are concerned with here, are normally assumed to be well-mixed axially. If the physical properties of the bed particles vary, however, the denser and/or larger particles will tend to segregate to the bottom (act as "jetsam"). Particle density is more important than size in determining which particles, e.g. in a binary mixture, will act as jetsam.

3.2 Bubbling Beds

In the following sections we will describe a stochastic model for the particle transport in a bubbling bed and its validation with experiment. The model formulation is somewhat a-typical in that it does not progress straight from the phenomenological description of the particle motion to the transfer probability matrix, but takes a de-tour through a description of the particle transport by continuous processes. This is due to the tradition in the fluidization literature of quantifying the particle transport mechanisms by continuous process parameters, such as particle superficial velocities and dispersion coefficients, as we shall see below.

For this reason we would like to point out at this stage that the application of stochastic modeling to a slugging bed described later, in Section 3.3, is probably better suited as a "text-book" example of stochastic model formulation than the bubbling bed model described here.

3.2.1 Particle Transport Mechanisms and their Quantification

Rowe and co-workers [50, 108] were the first to propose that the vertical particle motion in batch fluidized beds is governed by the following phenomena (see also Figure 3.2):

1. Transport upward in the wakes of fluidization bubbles and deposition on the bed surface.

2. Transport down in the bulk to compensate for this (1 and 2 together are termed 'circulation').

3. Dispersion due to disturbance of the bed material by fluidization bubbles.

The model presented here is based on these transport processes. We note that in many practical systems, particularly large beds, an uneven bubble distribution over the bed cross-section may spontaneously occur. In a localized region of strong bubble flow some particles between the bubbles may well be transported upward in addition to the transport in bubble wakes. This phenomenon is called "gulf streaming", and may lead to much higher particle circulation than if transport was in wakes only. If gulf streaming can be quantified, there is no mathematical difficulty in incorporating it in the model described below. We will be discussing this issue in Chapter 8.

Particle segregation normally takes place in a given region of the bulk material only if this region is being disturbed by a fluidization bubble passing in the vicinity [58, 116].

The total volumetric flow in the empty part of fluidization bubbles Q_b, which when calculated per unit cross-sectional bed area is the superficial velocity of the gas flowing through the bed in the "bubble phase", is approximately given by the well-known "two-phase theory" [120]:

$$Q_b = A(U - U_{mf}) \tag{3.1}$$

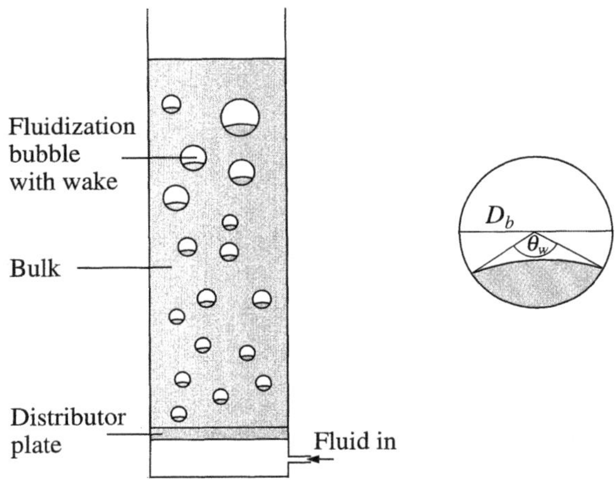

Figure 3.2: Bubbling batch fluidized bed. To the right a detail of a fluidization bubble and its associated wake

where A the the bed cross-sectional area. This implies that in a bubbling bed the superficial velocity of the gas percolating through the bulk particles remains U_{mf}.

The wakes of fluidization bubbles are normally assumed to the "sphere-completing", so that the fraction of the bubble-wake sphere filled with wake material, the "wake fraction", f_w, can be found from the angle subtended by the wake, θ_w, (see Figure 3.2) and knowledge of the radius of curvature of the upper wake surface. Empirical information shows that, the latter can be taken as approximately 2.3 times the radius of curvature of the bubble front, and the former, in degrees, as [19]:

$$\theta_w = 160 - 160\exp(-60D_b). \tag{3.2}$$

Calculating the wake fraction from the geometry and fitting the result of this with a simple functional form results in the following expression for the wake fraction [19]:

$$f_w = 0.45(1 - e^{-60D_b})^{2.75} \tag{3.3}$$

The wake fraction thus increases with bubble size, D_b, and we require D_b as a function of height, h, in the bed. A number of empirical formulae have been proposed for this [4], one is [46]:

$$D_b = \frac{1.3}{g^{0.2}}\left(\frac{U - U_{mf}}{1000}\right)^{0.4}\left(\frac{1}{1 - f_w}\right)^{0.33} + 2.05h\left(U - U_{mf}\right)^{0.94}, \quad (3.4)$$

where g is the gravitational acceleration. The divisor 1000 is for a porous distributor plate. The first term on the right-hand-side, which is due to Kato and Wen [74], gives the size on formation at the gas distributor, and the second the growth with height in the bed due to coalescence. A slightly later—and many say more accurate—empirical model is that of Mori and Wen [92]:

$$D_b = D_{b,m} - (D_{b,m} - D_{b,0})\exp\left(-0.3\frac{h}{D_{bed}}\right), \quad (3.5)$$

where the bubble diameter upon formation at the gas distributor plate, $D_{b,0}$, and the maximum achievable bubble size in the given bed, $D_{b,m}$, are given by, respectively:

$$\begin{aligned} D_{b,0} &= \frac{1.38}{g^{0.2}}\left(\frac{1}{1000}(U - U_{mf})\right)^{0.4} \\ D_{b,m} &= 1.49\left(D_{bed}^2(U - U_{mf})\right)^{0.4} \end{aligned} \quad (3.6)$$

D_{bed} is the diameter of the column. Note that the two constants in these expressions are dimensional, SI units have been used.

Since the bubble size thus increases with height in the bed, and the wake fraction with the bubble size, an increasing amount of material is transported in wakes as one moves up the bed, in spite of the fact that the two-phase theory implies that the flow of gas in the bubble voids is constant over the height of the bed. Thus there is a non-zero probability of bed material being caught up in wakes everywhere in the bed. A significant fraction of the wake material is, however, caught up in bubble wakes at the distributor plate, where bubbles and their associated wakes are formed.

The dispersion and segregation in the bulk material caused by the passage of one fluidization bubble can be estimated from the work of Tanimoto et al. [116], who determined the drift of particles, initially placed in a horizontal layer, after the passage of a single bubble (see Figure 3.3).

We call the coordinates of the Tanimoto plot, which are scaled with the radius of the bubble void's volume-equivalent sphere, x and y, the area of the bed (unity) A, the number of particles per unit bed area n_p, the bubble diameter D_b and the diameter of the bubble's volume-equivalent sphere $D_{b,e}$. We further call the physical particle displacement $\xi = y(D_{b,e}/2)$, and the particle's physical radial distance from the bubble centre $\zeta = x(D_{b,e}/2)$.

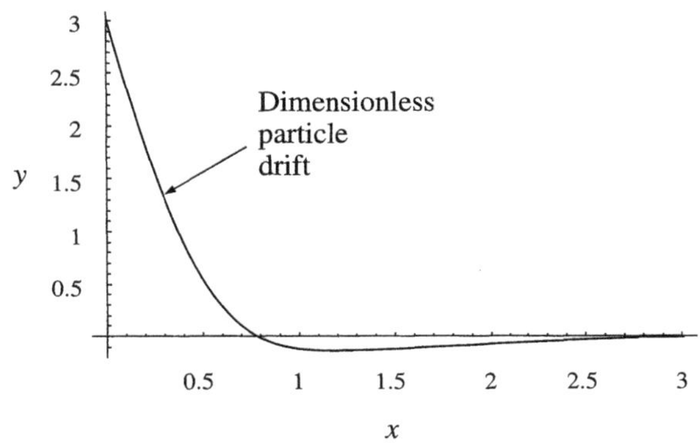

Figure 3.3: Sketch of the drift profile for particles, initially positioned in a horizontal layer, due to the passage of one fluidization bubble according to Tanimoto et al. [116]. Both the vertical displacement and the distance from the bubble center have been reduced by the radius of a sphere of the same volume as the empty part of the fluidization bubble

The total squared displacement of the particles in an differential element at radius ζ and thickness $d\zeta$ is then:

$$n_p 2\pi\zeta d\zeta\xi^2.$$

The total squared displacement due to the passage of one bubble becomes:

$$\int_{\zeta=0}^{\zeta=2.45\frac{D_{b,e}}{2} \text{ or } \infty} n_p 2\pi\zeta\xi^2 d\zeta$$

where we have indicated that the upper bound of the integral could also have been taken as ∞.

Changing the variable of integration to $x = \frac{\zeta}{D_{b,e}/2}$ such that $\frac{d\zeta}{dx} = \frac{D_{b,e}}{2} \Rightarrow d\zeta = \frac{D_{b,e}}{2}dx$ and introducing the new variable $y = \frac{\xi}{D_{b,e}/2}$:

$$\int_{x=0}^{x=2.45} n_p 2\pi x \frac{D_{b,e}}{2}\frac{D_{b,e}}{2}\left(y\frac{D_{b,e}}{2}\right)^2 dx.$$

Then the mean squared displacement is obtained by dividing by the total number of particles, which is n_p in unit area of bed:

$$\frac{1}{n_p}\int_{x=0}^{x=2.45} n_p 2\pi x \frac{D_{b,e}}{2}\frac{D_{b,e}}{2}\left(y\frac{D_{b,e}}{2}\right)^2 dx = \int_{x=0}^{x=2.45} 2\pi x y^2 \left(\frac{D_{b,e}}{2}\right)^4 dx$$

The mean squared displacement is linear in time, so to obtain the mean squared displacement per second we multiply by the number of bubbles crossing unit area of bed per second, n_b:

$$n_b = \frac{(U - U_{mf})}{\frac{\pi}{6}D_b^3(1 - f_w)}$$

to obtain:

$$\int_{x=0}^{x=2.45} 2\pi x y^2 \left(\frac{D_{b,e}}{2}\right)^4 dx \frac{(U - U_{mf})}{\frac{\pi}{6}D_b^3(1 - f_w)} \tag{3.7}$$

Now:

$$\left.\begin{array}{l} V_{b,e} = \frac{\pi}{6}D_b^3(1 - f_w) \\ V_{b,e} = \frac{\pi}{6}D_{b,e}^3 \end{array}\right\} \Rightarrow D_{b,e}^3 = D_b^3(1 - f_w)$$

where $V_{b,e}$ is the volume of the empty part of the fluidization bubble. Insert this in Equation (3.7):

$$\int_{x=0}^{x=2.45} 2\pi x y^2 \frac{D_b^3}{2^4} D_b(1 - f_w)^{\frac{1}{3}} dx \frac{(U - U_{mf})}{\frac{\pi}{6}D_b^3(1 - f_w)}$$
$$= \frac{3}{8\pi} D_b(1 - f_w)^{\frac{1}{3}}(U - U_{mf})1.75 = 0.2089 D_b(1 - f_w)^{\frac{1}{3}}(U - U_{mf})$$

where we have used that a numerical evaluation of the integral comes to about 1.6 (note that this is a very approximate value due to the scatter in the results of Tanimoto et al.) We thus get:

$$\text{Mean squared displacement} = 2D = 0.2089 D_b(1 - f_w)^{\frac{1}{3}}(U - U_{mf}) \tag{3.8}$$

where D is the Fickian dispersion coefficient.

This expression is slightly different from that given in [35], but is more consistent with the definitions of Tanimoto et al.; the correction is for formal reasons and does not change the predictions of the model below visibly. Note also that in [35] D is defined as the mean square particle displacement per second, while D in this book is *half* the mean square displacement per second. This latter definition is consistent with the D in Fick's laws of diffusion, Equations (1.2) and (1.27).

Tanimoto et al. [116] expressed the segregation of jetsam particles as a "segregation distance", $\bar{Y}_s$, which is the average downward displacement, again reduced by the radius of a sphere with the same volume as the empty part of the bubble: $D_{b,e}/2$, of the jetsam particles within the horizontally projected area of the volume-equivalent bubble, $\pi D_{b,e}^2/4$, caused by the passage of one fluidization bubble. Their empirical expression was modified later [19], to give:

$$\bar{Y}_s = 0.8 \left[\frac{\rho_j d_j^{0.33}}{c_f(\rho_f d_f^{0.33}) + (1 - c_f)(\rho_j d_j^{0.33})} - 1 \right] \tag{3.9}$$

where ρ and d are the particle density and diameter, respectively, c is the volumetric fraction, and sub-j and sub-f signify the jetsam and the flotsam particles, the latter being those particles that tend to rise in the bed.

Similarly to deriving a continuous parameter, D, to describe the dispersion process, the segregation can be averaged over time and over the bed cross-sectional area and expressed as a superficial velocity, v_{seg}, of jetsam particles through the bed. This is described in Section 3.2.5 below.

For many of the above expressions, we require the minimum fluidization velocity, U_{mf}, of the fluidized particles. This can, for particles of a uniform density, ρ, be estimated by the expression of Wen and Yu [132]:

$$\frac{d_p U_{mf} \rho_g}{\mu} = \sqrt{33.7^2 + 0.0408 \frac{d_p^3 \rho_g (\rho - \rho_g) g}{\mu^2}} - 33.7 \tag{3.10}$$

where μ and ρ_g are the gas viscosity and density, respectively. For binary mixtures of powders an empirical relation for the minimal fluidization velocity of the mixture $U_{mf,m}$ is [23]:

$$U_{mf,m} = U_{mf,f} \left(\frac{U_{mf,j}}{U_{mf,f}} \right)^{x_j^2}, \tag{3.11}$$

where x_j is the volume fraction of jetsam particles.

This is sufficient to quantify the particle transport processes in bubbling fluidized beds.

3.2.2 Model Formulation

The particle motion is modeled as a convection-diffusion process with segregation, modified by allowing jumps upward due to transport in bubble wakes. To make the model more general, and account for some additional industrial fluidized bed systems, we allow for a series of horizontal sieve-like baffles in the bed, that knock out some of the wake from the bubbles as they rise through the bed. Such a feature may also simulate exchange of material between the bubble wake and the surrounding bulk, an exchange that in the literature is assumed to take place in many systems.

We consider a bed of unit cross-sectional area so that we can equate superficial velocities and volumetric flows. We can do this without loss of generality, since we assume all transport processes to be cross-sectionally uniform.

The motion of one particle is considered, and the transport processes are converted to transition probabilities between cells in a discretized bed, see Figure 3.4. The probability distribution for the particle's position as a function of time reflects the behavior of a pulse of marked particles.

We model the motion of a single particle in the bed by a stochastic process $(X_t)_{t\geq 0}$, where X_t gives the vertical distance of the particle from the bed surface at time t, and we assume that the process is Markovian. In our discretized bed, subdivided into a finite number of cells C_i, $1 \leq i \leq N$ and modeling only transitions at times $n \geq 0$, we obtain a Markov chain $(X_n)_{n\geq 0}$ with state space $\{1, \ldots, N\}$. The Markov chain is specified by its transition probabilities $p_{ij}(n) := P(X_{n+1} = j | X_n = i)$. As can be gleaned from what went before, the bubbling fluidized bed process can be considered homogeneous in time, making the transition probabilities independent of time, so that the argument of the p_{ij} can be dropped. Later in this chapter we will consider a model for a slugging bed, a process that we will, initially at least, not treat as homogeneous in time.

The cells are numbered as shown in Figure 3.4. The model calculates the probability distribution of the axial position of one particle as a function of time. The possible transitions are:

1. staying in the same cell
2. moving to the next cell
3. moving back to the previous cell
4. being caught up in a bubble wake and deposited under one of the baffles passed by the rising bubble, or at the top of the bed. We assume that only part of the wake is retained by a baffle

We introduce parameters α_i, β_i and δ_i, with sum equal to 1, for the first three probabilities, conditionally on the particle not being caught up in a bubble wake, the latter probability being given by λ_i.

The transfer probabilities from cell i to cell j form a matrix, $\mathbf{P}$, with the elements p_{ij}.

Let us for illustration first write down the transfer probability matrix for the case where no baffles are present in the bed. In that case, a particle caught up in a bubble wake is always deposited on the bed surface. For this case, the transition probabilities for the interior of the bed, i.e. for

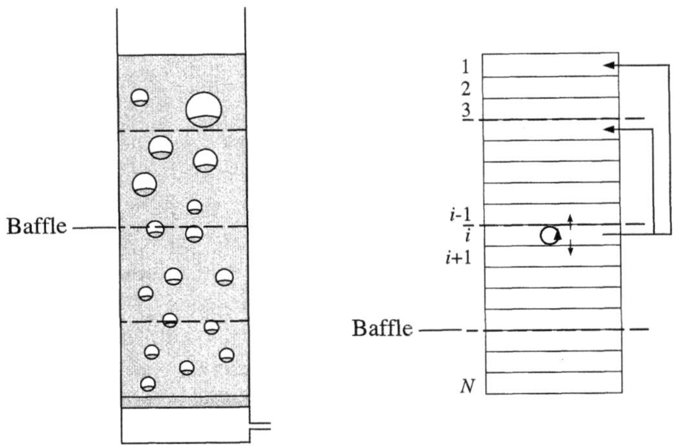

Figure 3.4: Diagram of a bubbling fluidized bed with horizontal baffles, and the discretized fluidized bed with baffles

$2 \leq i \leq N$, are:

$$\begin{aligned} p_{i,i} &= \alpha_i(1-\lambda_i), \\ p_{i,i+1} &= \beta_i(1-\lambda_i), \\ p_{i,i-1} &= \delta_i(1-\lambda_i), \\ p_{i,1} &= \lambda_i. \end{aligned} \tag{3.12}$$

Regarding the boundaries, i.e., $i = 1$ and $i = N$:

$$\begin{aligned} p_{1,1} &= 1-\beta_1(1-\lambda_1), \\ p_{1,2} &= \beta_1(1-\lambda_1), \\ p_{N,N} &= 1-\delta_N(1-\lambda_N)-\lambda_N. \end{aligned} \tag{3.13}$$

Writing out the matrix for the case of no baffles in the bed:

$$\begin{pmatrix} 1-\beta_1(1-\lambda_1) & \beta_1(1-\lambda_1) & 0 & \dots & 0 & 0 \\ \lambda_2+\delta_2(1-\lambda_2) & \alpha_2(1-\lambda_2) & \beta_2(1-\lambda_2) & \dots & 0 & 0 \\ \lambda_3 & \delta_3(1-\lambda_3) & \alpha_3(1-\lambda_3) & \dots & 0 & 0 \\ \lambda_4 & 0 & \delta_4(1-\lambda_4) & \dots & 0 & 0 \\ \vdots & \vdots & \vdots & \ddots & \vdots & \vdots \\ \lambda_N & 0 & 0 & \dots & \delta_N(1-\lambda_N) & 1-\delta_N(1-\lambda_N)-\lambda_N \end{pmatrix}$$

The matrix is sparse, except for the non-zero elements in the first column; it has a lower tridiagonal structure. This feature makes evaluation of the matrix very advantageous in terms of computational effort.

We now consider the case where a number of baffles are inserted in the bed, and denote by θ the fraction of wake retained by each baffle. Since some material caught in wakes is deposited under the baffles, the probability of return to the top of the bed, $p_{i,1}$ is reduced, and new non-zero probabilities are introduced for transfer to the cells just under each baffle, giving rise to a matrix with more off-diagonal non-zero elements:

$$\begin{aligned} p_{i,1} &= \lambda_i (1-\theta)^{\left[\frac{i-1}{S}\right]}, \\ p_{i,mS+1} &= \lambda_i \theta (1-\theta)^{\left[\frac{i-1}{S}\right]-m}; \qquad m = 1, 2, \ldots, \left[\frac{i-1}{S}\right]. \end{aligned} \tag{3.14}$$

Note that $\lambda_i = \lambda_i\theta + \lambda_i\theta(1-\theta)^1 + \ldots + \lambda_i\theta(1-\theta)^{\left[\frac{i-1}{S}\right]-m} + \ldots + \lambda_i(1-\theta)^{\left[\frac{i-1}{S}\right]}$. m is the index of the baffles and S is number of cells between two baffles. The square brackets indicates "the integer part of...".

All the other components of the transfer probability matrix remain the same as for the case with no baffles.

The position of the particle at the n'th time step is given by the probability vector $\mathbf{p}(n)$ with elements $p(n, i)$.

Knowing $\mathbf{p}(n-1)$, one can find $\mathbf{p}(n)$ from the recursion formula:

$$p(n, j) = \sum_{i=1}^{N+1} p(n-1, i) p_{ij}$$

or in matrix notation:

$$\mathbf{p}(n) = \mathbf{p}(n-1)\mathbf{P}. \tag{3.15}$$

After n time steps, we obtain the formula for the probability distribution of position of the particle at time n in terms of its initial probability distribution:

$$\mathbf{p}(n) = \mathbf{p}(0)\mathbf{P}^n \tag{3.16}$$

where $\mathbf{p}(0)$ is the initial condition of particle distribution in the reactor at time $t = 0$.

3.2.3 Relating the Transfer Probabilities to the Physical Processes

The model introduced above is a discrete one, but the transfer probabilities will, as mentioned, be related to physical parameters describing the particle transport as continuous processes, following Dehling et al. [39]. We call the time step ε and the cell width Δ. Letting ε and Δ go to 0, we obtain a discrete Markov chain approximation to a continuous process.

In this model formulation, in contrast to some other ones later in this book, we choose to maintain the physical parameters in their dimensional form rather than form dimensionless parameters by using characteristic parameters for scaling.

The vertical distance from the top of the reactor is denoted by x, i.e., $x = 0$ corresponds to the top and $x = H$, where H is the total height of the bed, to the bottom, and the convective axial velocity due to circulation by $v_{cir}(x)$. The dispersion due to the disturbance by bubbles is described by a dispersion coefficient, $D(x)$. The rate of returns to one of the baffles or to the top of the bed is described by $\lambda(x)$. The parameters in the transition matrix are defined as follows:

$$\begin{aligned}
\delta_i &= \frac{\varepsilon}{\Delta^2} D(i\Delta) - \frac{\varepsilon}{2\Delta} v_{cir}(i\Delta), \\
\beta_i &= \frac{\varepsilon}{\Delta^2} D(i\Delta) + \frac{\varepsilon}{2\Delta} v_{cir}(i\Delta), \\
\alpha_i &= 1 - \delta_i - \beta_i, \\
\lambda_i &= \varepsilon \lambda(i\Delta).
\end{aligned} \tag{3.17}$$

With these choices of transfer probabilities, conditioned on the particle not being caught up in a bubble wake, the mean displacement per step becomes $\varepsilon v_{cir}(i\Delta)$, which corresponds to a mean velocity of $v_{cir}(i\Delta)$. The mean squared displacement per step is given by $\Delta^2(\beta_i + \delta_i) = 2\varepsilon D(i\Delta)$, resulting in a mean squared displacement per unit of time of $2D(i\Delta)$.

Finally λ_i, the probability that a particle is caught in a bubble wake, is calculated as the fraction of material in a cell caught up in bubble wakes during one time step:

$$\lambda_i = \frac{v_{cir}((i-1)\Delta) - v_{cir}(i\Delta)}{\Delta} \varepsilon. \tag{3.18}$$

For cell N, $\lambda_N = (v_{cir}(N\Delta)/\Delta)\varepsilon$.

3.2.4 Modeling of Segregation

The model outlined so far does not account for segregation. We model only the jetsam fraction. Since segregation adds an extra convective downward velocity, a simple way of modeling this is to modify the above probabilities as follows:

$$\begin{aligned}\delta_i^* &= \delta_i, \\ \beta_i^* &= \beta_i + \frac{\varepsilon}{\Delta} v_{seg}(i\Delta), \\ \alpha_i^* &= 1 - \delta_i^* - \beta_i^*,\end{aligned} \tag{3.19}$$

λ_i^* equals λ_i, and v_{seg} is the superficial velocity of jetsam segregation. These, then, are the final transition probabilities in the interior, i.e. for $i = 1, \ldots, N$.

3.2.5 Quantification of the Model Parameters

We now quantify the parameters D, v_{cir} and v_{seg}, using the empirical relationships given above, remembering that we consider unit cross-sectional bed area, and can therefore equate superficial velocities and volumetric flows.

We have already discussed the dispersion coefficient D, which is a function of height in the bed. This was computed by calculating the total squared displacement caused by one fluidization bubble from the profiles of Tanimoto et al. [116], accumulating this over one second, using Equations (3.1) and (3.4) to calculate the bubble frequency and size, and calculating the mean squared particle displacement during one second, which is equal to $2D$.

v_{cir} is the superficial velocity of material transported in bubble wakes, and is calculated as:

$$v_{cir} = \frac{(U - U_{mf})f_w}{1 - f_w}, \tag{3.20}$$

since it follows from Equation (3.1) that the superficial velocity of the bubble-wake sphere is $(U - U_{mf})/(1 - f_w)$.

v_{seg} is calculated as:

$$v_{seg} = \bar{Y}_s \frac{D_{b,e}}{2} F_b \tag{3.21}$$

where F_b is the point bubble frequency, which is given by:

$$F_b = (U - U_{mf}) \frac{A_{b,e}}{V_{b,e}} = (U - U_{mf}) \frac{3}{2D_{b,e}}. \tag{3.22}$$

We have written these two equations in terms of the diameter, the projected area and the volume of a sphere with the same volume as the empty part of the bubble, or "volume-equivalent sphere", $D_{b,e}$, $A_{b,e}$ and $V_{b,e}$, since this, as mentioned above, is what Tanimoto et al. used as a scaling factor for $\bar{Y}_s$.

3.2.6 Numerical Simulations and Experimental Validation

Experiments to validate the model are described in detail in [35]. The experiments were carried out in a glass column of 15 cm diameter equipped with baffles, and containing a binary mixture, 50/50 by volume, of glass beads (83 μm) and painted glass beads (221 μm). The properties of the solids are shown in Table 3.1.

Table 3.1: Properties of the particles used

Particles	d_p (μm)	σ^* (-)	ρ (kg/m^3)	U_{mf} (cm/s)	ε_{mf} (-)
Glass beads	83	0.11	2500	0.595	0.416
Painted glass beads	221	0.09	2480	5.290	0.427

*The standard deviation of a log-normal distribution fitted to the particle size distribution

The baffles used consisted of woven wires of 0.65 mm diameter with stitch of 0.42 cm giving 71.1% open area. The spacing between the baffles varied between 0.43 and 7.40 cm.

The experiments were started at a high fluidization velocity at which the bed was well mixed. The experimental conditions were then set and the segregation allowed to reach equilibrium. Sampling was done by sectioning the bed in horizontal slices with a vacuum technique after having defluidized the bed suddenly to freeze the composition.

To compare the experimental data with the model the probabilities $p(n, i)$ were converted to jetsam concentration $c(n, i)$ using:

$$c(n, i) = NC_j p(n, i),$$

where C_j is the volumetric jetsam concentration in the entire bed. Obviously the physics imposes a maximum concentration of $c(n, i) = 1$. This is not yet accounted for in the model, and was imposed in the numerical evaluation of the model. We address the issue of modeling interfering particles in Section 3.4.

All of the experimental results used for comparison were obtained using a superficial fluidization velocity, U, higher than the U_{mf} of the jetsam to avoid defluidization in the bottom of the bed. Figures 3.5–3.7 show the experimental results as points. Clearly, the more baffles used, the better the segregation of jetsam.

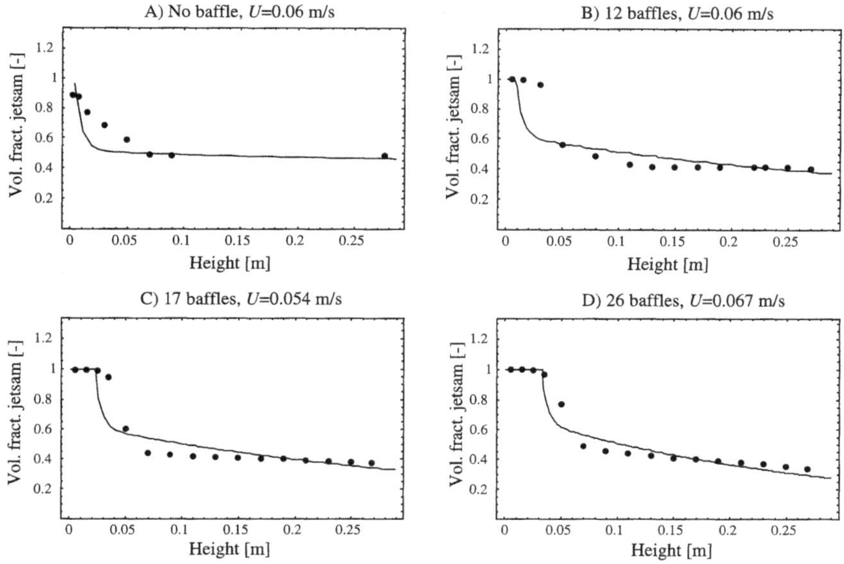

Figure 3.5: Comparison between experimental data (points) and the stochastic model (lines). Baffle separations are: A) No baffles, B) 2.40 cm, C) 1.72 cm and D) 1.13 cm

There is one adjustable parameter in the model, θ, the fraction of wake left under the baffle. In all of the plots shown the same value of θ has been used, namely 0.03. Even so, the agreement between model and experiment is clearly good. The only effect not well accounted for is that of the fluidization velocity. The model correctly reflects:

- the effect of the number of baffles on the separation without having to adjust θ,
- the effect of the baffles in causing an effective bulk/wake "exchange", which gives rise to a gradient in the jetsam concentration in the upper part of the bed.

The numerical value of 0.03 for θ is very low, especially in view of the X-ray pictures of van Dijk et al. [125], which show most of the wake material left under a baffle by a single fluidization bubble. In the work of van Dijk et al., however, the wake material had a significantly higher density than

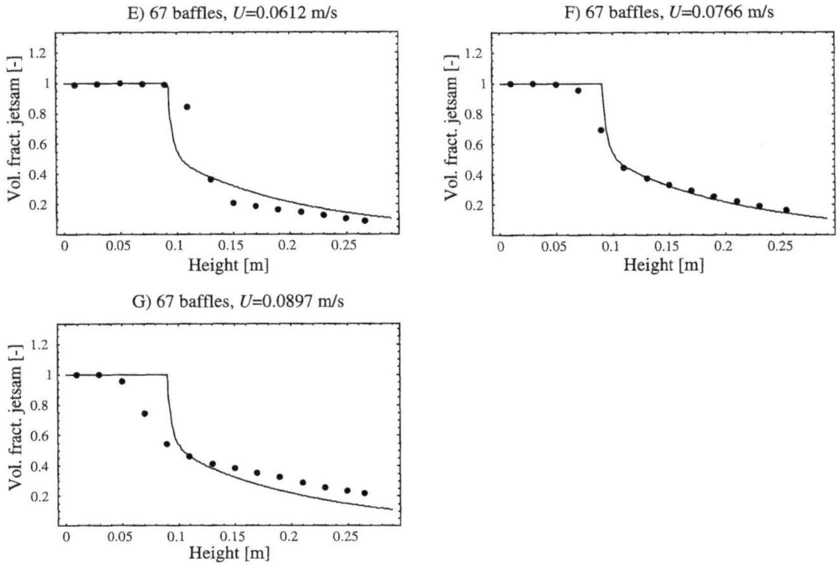

Figure 3.6: Comparison between experimental data (points) and the stochastic model (lines) for the same number of baffles (baffle distance = 0.43 cm) but three different superficial fluidization velocities

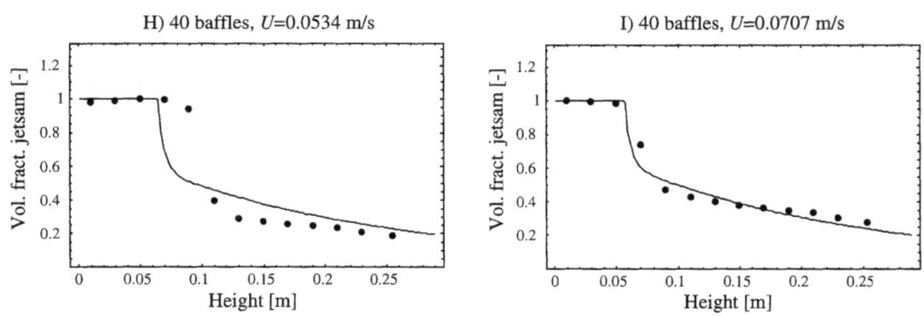

Figure 3.7: Comparison between experimental data (points) and the stochastic model (lines) for the same number of baffles (baffle distance = 0.72 cm) and two different superficial fluidization velocities

the surrounding bulk material, and it is possible that the wake therefore was more easily left under the baffle. The relative size of the fluidization bubble and the baffle aperture may also play a role in determining the baffle retention efficiency.

The agreement between model and experiment seen in the figures show that the stochastic model, which can be evaluated in a PC in a few seconds, is capable of predicting the profile of segregation in baffled batch fluidized bed with a binary mixture as discussed above.

3.3 Slugging Fluidized Beds

In this section we will formulate a stochastic model for the particle transport in a slugging fluidized bed; a system so complex that the normal modeling approach of formulating differential conservation equations fails. At the same time, this is an excellent example of stochastic modeling since we go directly from the particle transport processes to the model without taking a detour around continuous parameter definitions, as we were forced to do in the previous example for bubbling beds by the need to use the existing literature to quantify the model parameters.

We note at this point that the system we model in this chapter and in Chapter 5, namely a fluidized bed, often, but not always, contains so many particles that the process becomes deterministic in nature. We stress that this does not detract from the advantages of using a stochastic modeling approach: advantages in the model formulation and also advantages in the model solution, as we shall see particularly clearly in this section.

3.3.1 Particle Transport Mechanisms and Their Quantification

fluidization normally occurs in beds with a high height-to-diameter ratio. A slug is simply a fluidization bubble, the diameter of which is nearly equal to that of the bed itself. Even so, slugs influence the bed material in a different way than smaller bubbles do. Some research literature is dedicated to slug type fluidization, though it is still relatively scarce. Extensive empirical information about the nature of slugging fluidized beds is available in references [32, 90, 112]. Abanades and Atarés [1] examined slugging fluidization in a British Coal pressurized fluidized bed gasifier. We shall be comparing the predictions of our model with their experimental results. Figure 3.8 illustrates two modes of slugging observed in fluidized beds. In this section we focus on axisymmetric slugging, although the modeling approach equally well can be used for fluidized beds with wall slugs.

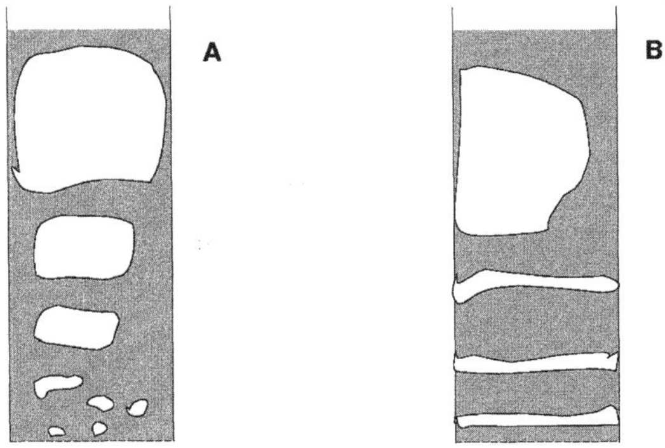

Figure 3.8: Modes of slugging: **A** axisymmetric and **B** wall slugs [3]

Again we are interested in a binary mixture of particles in which a fraction segregates to the bottom in competition with mixing caused by rising slugs.

We have mentioned the tendency of denser and/or larger particles to segregate to the bottom of a fluidized bed. Binary mixtures have often been used for studies of this phenomenon. Binary mixtures are usually referred to as "equal-density mixtures" if the two fractions differ in size only, and "different-density mixtures" if the density or both properties are different [50, 63]. However, as mentioned, the difference in density tends, if present, to be the dominant driving force for segregation, and this is our interest here. The development of a model for mixing and segregation in bubbling fluidized beds was pioneered by Rowe and co-workers [50, 108], who considered that solids mixing and segregation are driven by the motion of fluidization bubbles. These were the concepts used for a model for mixing and segregation in bubbling batch fluidized beds in the previous section, and by Dehling et al. [39] for a stochastic model for particle transport in continuous beds, which will be discussed in a subsequent chapter of this book.

3.3.2 Model Formulation

Again, we model the motion of a single particle in the reactor by a stochastic process $(X_t)_{t\geq 0}$, where X_t gives the location of the particle at time t, and we assume that the process is Markovian. Like for the bubbling bed in the previous section, we discretize time and space, subdividing the fluidized bed into a finite number of cells C_i, $1 \leq i \leq N$ and modeling only transitions at times $n \geq 0$. We thus obtain a Markov chain $(X_n)_{n \geq 0}$ with state space $\{1, \ldots, N\}$ and the transition probabilities $p_{ij}(n) := P(X_{n+1} = j | X_n = i)$.

Given an initial distribution vector $(\mathbf{p}(i))_{1 \leq i \leq N}$ and the transition matrices $\mathbf{P}(n) := (p_{ij}(n))_{1 \leq i,j \leq N}$ for all $n \geq 0$ the probability distribution of the particle at some time $n_0 \geq 0$ is can easily be calculated via

$$(P(X_{n_0} = i))_{1 \leq i \leq N} = \mathbf{p}\mathbf{P}(1) \cdot \ldots \cdot \mathbf{P}(n_0). \tag{3.23}$$

3.3.3 A Stochastic Model for a Slugging Fluidized Bed

We consider a fluidized bed with height at rest h_b, as shown in Figure 3.9.

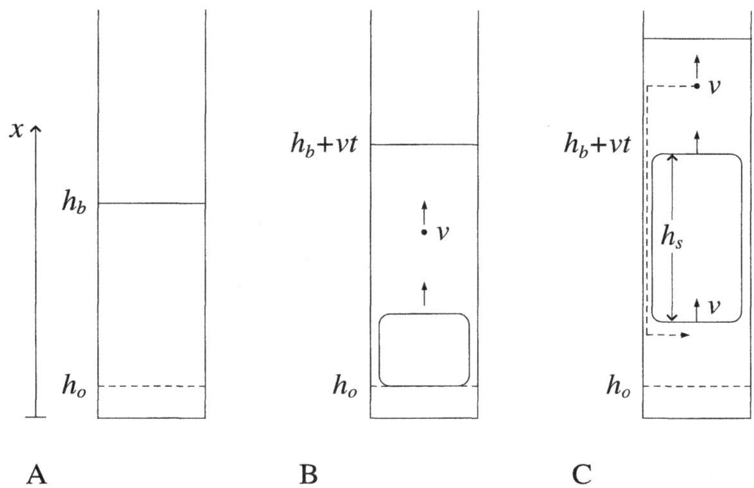

Figure 3.9: Slugging fluidized bed in different states of development: **A** Bed at rest, **B** Slug formation phase and **C** Slug rise phase.(The dot represents a single particle)

As shown in reference [32], slugs are formed by at some height h_0 in the bed by coalescing fluidization bubbles. They grow in size until they have reached a certain height h_s at which time they 'detach'[2] and start rising as

[2]'Detachment' actually means that a new slug starts to form underneath the slug in question, but in our model concept only one slug is present in the bed at a time

a whole toward the top of the bed. When the nose of the slug has reached the top of the bed, the slug collapses. We can thus distinguish two separate phases, as follows:

Slug formation: The slug forms at a height h_0 above the bottom of the bed and grows in vertical size at some speed from 0 to h_s. In this phase, all particles above the slug are pushed upward, and as a result the bed height also rises above its original level h_b. Assuming that the slugs nearly fill up the entire diameter of the reactor, the bed also rises at the same speed as the slug to its maximum height $H := h_b + h_s$ (see Figure 3.9).

Slug rise: When the slug has reached its final height, it starts rising toward the top of the bed. In this phase the height of the slug, and consequently also the height of the bed, remain constant. At the end of the slug rise phase, the nose of the slug has reached the top of the bed. At this point the slug ceases to exist, and the bed collapses back to its original height.

As a result of slug formation and rise particles in the bed are displaced. They are either pushed upward or fall through the annulus around the slug and then are deposited directly underneath the slug. The probability of the particle moving to the annulus and down below the slug is likely to depend on the vertical distance of the particle from the nose of the slug, x and its rate is a function, $\lambda_h(x)$, of h where h denotes the height of the slug nose and $h \leq x \leq H$. In accordance with [32] we envisage the flow pattern as sketched in Figure 3.10.

We consider a bed of unit cross-sectional area so that we can equate superficial velocities and volumetric flows.

The key concepts behind the model are as follows. The effect of formation and rise of a single slug are modeled first. The displacement of particles due to several consecutive slugs, even if more than one is present in the bed at any one time, is assumed to be a superposition of the effects of the individual slugs. On this basis, a model for the displacement of the bulk material in time is formulated; whereafter the segregation of the particle fraction tending to sink is considered in a subsequent section. To simplify the model, the horizontal component of particle displacement is neglected and only vertical displacement is considered. According to Davidson and Harrison [32] the rate function $\lambda_h(x)$ is a decreasing function of x, i.e., more of the particles that deposit underneath the slug come from close to the slug nose than from the top of the bed. The slug is assumed to move upward at a constant speed v. The total flow around the slug must equal the speed at which the slug rises; therefore we obtain the continuity condition

$$\int_h^H \lambda_h(x)dx = v, \tag{3.24}$$

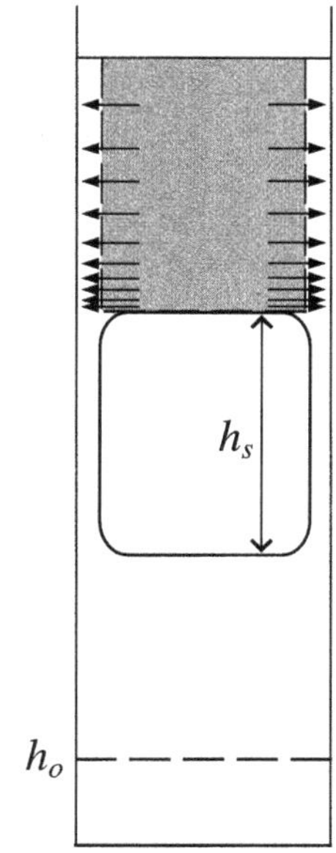

Figure 3.10: A sketch indicating the particle movement to the annulus in front of slug nose. Most of the particles that deposit underneath the slug come from the region close to the slug nose rather than from higher in the bed. The density of arrows indicate the magnitude of the particle flux

for any h in the interval $[h_s, H]$. In the following, it is assumed that $\lambda_h(x)$ follows a power law, that is $\lambda_h(x) = c(H + h - x)^r$ with $r \geq 1$, where the constant c is determined by the continuity condition (3.24)

$$\lambda_h(x) = \frac{v(r+1)}{H^{r+1} - h^{r+1}}(H + h - x)^r, \tag{3.25}$$

for all $h_s \leq h \leq H$ and $h \leq x \leq H$.

The exponent r still remains to be determined, either from experimental observations or from considerations concerning the flow of particles from the region above the slug via the annulus to the bottom part of the bed (see Figure 3.10).

Particles above the slug that do not flow into the annulus rise at a speed $v_h(x)$ which depends on the vertical location. It is computed as the difference between the speed of the slug nose and the flow of particles into the annulus below height x, that is

$$v_h(x) = v - \int_h^x \lambda_h(y)dy.$$

Using equation (3.24), this can also be written as

$$v_h(x) = \int_x^H \lambda_h(y)dy, \tag{3.26}$$

which can be interpreted as equating the upward flow of the rising particles to the downward flow in the annulus.

As mentioned, to obtain a discrete Markov chain approximation to the continuous system we divide the bed into horizontal slices of width Δ and discretize time into steps ε such that $\frac{\Delta}{\varepsilon} = v$. In this way, one step per unit of time corresponds to the speed v. Furthermore we assume that $h_0 = 0$ and that h_s and h_b are integer multipliers of Δ, i.e., $h_s = i_s\Delta$ and $h_b = i_b\Delta$. The horizontal slices (cells) are numbered by $i = 1, ..., N = i_s + i_b$. The discrete Markov chain model will have state space $\{1, \ldots, N\}$ and transition probabilities $p_{ij}(n)$, which depend on time and are defined below. Since $v = \frac{\Delta}{\varepsilon}$ it takes i_b time units for the formed slug (with its nose at height i_s already) to reach the top of the bed. Because in the slug formation phase (see section 3.3.1) all particles above the forming slug only move upward with velocity v we can assume that the slug has already formed by shifting the initial distribution $(\mathbf{p}(i))_{1\leq i\leq N}$ i_s cells upward to $\bar{\mathbf{p}} = \mathbf{pS}$ via the matrix $\mathbf{S} = (s_{ij})_{1\leq i,j\leq N}$ defined by

$$s_{ij} := \begin{cases} 1 & if\ 1 \leq i \leq i_b \text{ and } j = i + i_s\ , \\ 0 & \text{else.} \end{cases} \tag{3.27}$$

Therefore we start counting time after the slug has already formed, i.e., time 0 corresponding to the point of time just before the slug starts to rise. There are no transitions by particles below the slug nose. For the sake of completeness we set $p_{i,i}(n) = 1$ for all $i \leq n + i_s$ because at time n the slug nose is in cell $n + i_s$.

For a particle in a cell i above cell $n+i_s$ at time $n \leq i_b - 1$, i.e., $n+i_s \leq i \leq N$, there are three different possible transitions (see Figure 3.11)

- move to what will be the first cell below the slug after the time step, i.e., to $j = n$,
- move one cell up, i.e., to $j = i + 1$,
- stay in the same cell.

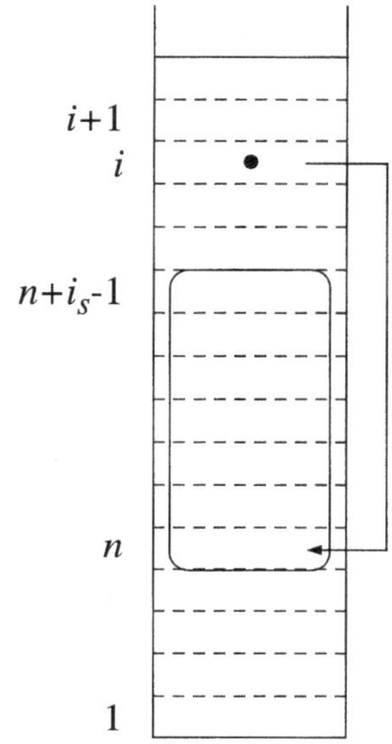

Figure 3.11: The discrete Markov chain model for the n-th transition. Just before the transition the slug occupies the cells $n, \ldots, n+i_s-1$. The arrow indicates the transition into the annulus, i.e. from cell i to cell n. The two other possible transitions are up into cell $i+1$ and staying in cell i

The probabilities for these three transitions can be deduced using the rate function $\lambda_h(x)$ (see equation (3.25)). The rate of particles moving from a cell into the annular region around the slug and being deposited directly underneath the slug at cell n is given as the average probability rate for removal from that cell multiplied by one time unit ε and calculated below. The upward movement can be derived with equation (3.26). We obtain

$$\begin{aligned} p_{i,n}(n) &= \varepsilon \frac{1}{\Delta} \int_{(i-1)\Delta}^{i\Delta} \lambda_{(n-1+i_s)\Delta}(x)dx \\ &= \frac{(H+(n-i+i_s)\Delta)^{r+1} - (H+(n-i-1+i_s)\Delta)^{r+1}}{H^{r+1} - ((n-1+i_s)\Delta)^{r+1}} \end{aligned} \tag{3.28}$$

$$
\begin{aligned}
p_{i,i+1}(n) &= \varepsilon\frac{1}{\Delta}\int_{i\Delta}^{H}\lambda_{(n-1+i_s)\Delta}(x)dx \\
&= \frac{(H+(n-i-1+i_s)\Delta)^{r+1}-((n-1+i_s)\Delta)^{r+1}}{H^{r+1}-((n-1+i_s)\Delta)^{r+1}}
\end{aligned}
\tag{3.29}
$$

$$
\begin{aligned}
p_{i,i}(n) &= 1-p_{i,n}(n)-p_{i,i+1}(n) \\
&= \varepsilon\frac{1}{\Delta}\int_{(n-1+i_s)\Delta}^{(i-1)\Delta}\lambda_{(n-1+i_s)\Delta}(x)dx \\
&= \frac{H^{r+1}-(H+(n-i+i_s)\Delta)^{r+1}}{H^{r+1}-((n-1+i_s)\Delta)^{r+1}}
\end{aligned}
\tag{3.30}
$$

for all $1 \leq n \leq i_b$ and $n+i_s \leq i \leq N$. Thus the transition matrices $\mathbf{P}(n) := (p_{ij}(n))_{1\leq i,j\leq N}$ are defined for all $1 \leq n \leq i_b$. The probability distribution at some time $0 \leq n_0 \leq i_b$ is now given by equation (3.23) as

$$\mathbf{p}(n_0) = \bar{\mathbf{p}}\mathbf{P}(1)\cdot\ldots\cdot\mathbf{P}(n_0), \tag{3.31}$$

where $\mathbf{p}(n_0) = (P(X_{n_0} = 1),\ldots,P(X_{n_0} = N))$. Considering $n_0 = i_b$ captures the total effect of one slug formation and rise. The probability distribution at time $n_0 = i_b$ is equal to the distribution after one slug of an infinite number of particles with initial distribution $\mathbf{p}$.

What we have modeled so far is the particle behavior in a bed containing particles of uniform properties. A simple segregation process is introduced based on the idea of "jetsam" particles (the heavier and/or larger particles that tend to sink in the mixture) sinking when slugs disturb the system. Segregation of jetsam and flotsam particles in a slugging fluidized bed manifests itself as a difference in the downward drifts. The mass balance equation (3.24) only holds on average, with $\int_h^H \lambda_h(x)dx > v$ for jetsam particles and $\int_h^H \lambda_h(x)dx < v$ for flotsam particles. The segregation effect is simulated by increasing the probability of a down-transition for the jetsam particles, giving an extra downward drift. A simple way to account for larger downward flow of the jetsam particles is to modify the transition probabilities (3.28), (3.29) and (3.30) as follows:

$$\bar{p}_{i,n}(n) = \alpha + (1-\alpha)p_{i,n}(n), \tag{3.32}$$

$$\bar{p}_{i,i+1}(n) = (1-\alpha)p_{i,i+1}(n), \tag{3.33}$$

$$\bar{p}_{i,i}(n) = (1-\alpha)p_{i,i}(n) \tag{3.34}$$

for all $1 \leq n \leq i_b$ and $n+i_s \leq i \leq N$. Here $0 \leq \alpha \leq 1$ denotes a constant defining the strength of the segregation effects. When calculating the probability distributions for jetsam particles the transition probabilities (3.32), (3.33) and (3.34) have to be used in equation (3.31).

The effect of several slugs is now calculated via superposition. Let $s \in \mathbb{N}$ be the number of slugs. Then for the probability distribution of the particle's position given its initial probability distribution $\mathbf{p}$ after s slugs it holds that

$$(P(X_s = i))_{1 \leq i \leq N} = \mathbf{p}(\mathbf{SP}(1) \cdot \ldots \cdot \mathbf{P}(i_b))^s \ , \tag{3.35}$$

with X_s denoting the jetsam particle's position after s slugs, $\mathbf{P}(i)$ for $1 \leq i \leq i_b$ defined by equations (3.32), (3.33), (3.34) and the shift matrix $\mathbf{S}$ defined by equation (3.27).

3.3.4 Model Results

The model predictions for Figure 3.12, 3.13 and 3.14 in Section 3.3.5 were calculated using MATLAB. Equation (3.35) was implemented with number of cells N between 15 and 17 ($N - 10$ of these were occupied by the slug), diameter of column $D_{bed} = 0.15$ m respectively $D_{bed} = 0.13$ m, minimum fluidization velocity $U_{mf} = 1$ m/s respectively $U_{mf} = 0.16$ m/s and height of bed h_b, segregation parameter α, parameter for the flow into the annulus region r, superficial gas velocity U, time t as given in the figures. The total tracer volume was given via the total tracer volume in the experimental data. It was assumed that slugging begins on the distributor plate and the slug frequency f can, following Abanades and Ararés [1], be calculated from the empirical correlation:

$$f = 0.32 \frac{U^{-0.15}}{h_b} \ .$$

All the physical properties were based on the axisymmetrical slug type. The height of a stable slug h_s was calculated by solving [31]

$$\frac{h_s}{D_{bed}} - 0.495 \sqrt{\frac{h_s}{D_{bed}}} \left(1 - \frac{U - U_{mf}}{0.35\sqrt{gD_{bed}}}\right) + 0.061 - \frac{1.939(U - U_{mf})}{0.35\sqrt{gD_{bed}}} = 0.$$

The calculations took only a few seconds on a standard PC.

3.3.5 Comparison of Simulation and Experiments

The experimental results and the results from our model are not in the same form. The experimental results are in terms of volume of tracer at some height in the bed as the fraction of the total volume of material, we call this $V_{fr,i}$. The model calculations, on the other hand, give p_i: the probability that the particle is in cell i, which can, in accordance with the law of large numbers, also be seen as the volume of tracer in cell i as a fraction of the total volume of tracer in the whole bed. To compare the two, the model calculations were converted to the same form as the experimental results. If the bed is discretized in N cells of equal volume, the volume fraction of each cell is $\frac{1}{N}$. If the total volume of tracer in the experimental bed is V_{tr}, then p_i can be converted to $V_{fr,i}$ as follows:

$$V_{fr,i} = \frac{V_{tr} p_i}{\frac{1}{N}} ,$$

for all $1 \leq i \leq N$.
In this section, we do not account for the effect of the jetsam particles filling up the cell yet. This would impose a maximum value for p_i of $p_{max} := p_{i,max} = \frac{1}{NV_{tr}}$ for all $1 \leq i \leq N$. However, to take full account of these effects in the model, the change of the segregation behavior of the jetsam particles in response to the change in the local jetsam concentration will also have to be accounted for. This issue will be considered separately in Section 3.4.

The two parameters r and α in the model are functions of the experimental conditions and particle properties. When sufficient information becomes available, these can be quantified directly. In this study we use them as adjustable parameters.

The experimental results are from Abanades and Atarés [1]. A layer of white jetsam particles was arranged in a bed of red flotsam particles. The solids used were pigment agglomerates. The mixture was suddenly fluidized at a pre-set gas velocity and the process recorded on videotape. The experiments were carried out in a column of 15 cm diameter and were performed at 3 superficial gas velocities, 1.57, 1.69 and 1.83 m/s, and 2 initial bed heights, 0.53 and 0.73 m.

Figure 3.12 shows plots of distribution of jetsam volume fraction as a function of height in the bed after 20 seconds for different bed heights and superficial gas velocities. The height in the bed at cell index c, h_c, is given as

$$h_c = \frac{c}{\text{Number of cells}} \times h_b .$$

The adjustable parameters are set to $r = 1$ and $\alpha = 0.17$ in all cases to give acceptable agreements between experimental results and model predictions.

Figure 3.12 shows good agreement between the predicted and experimental data. There is one weak effect in the experimental data that is not accounted for in the model predictions. The tracer fraction in the upper

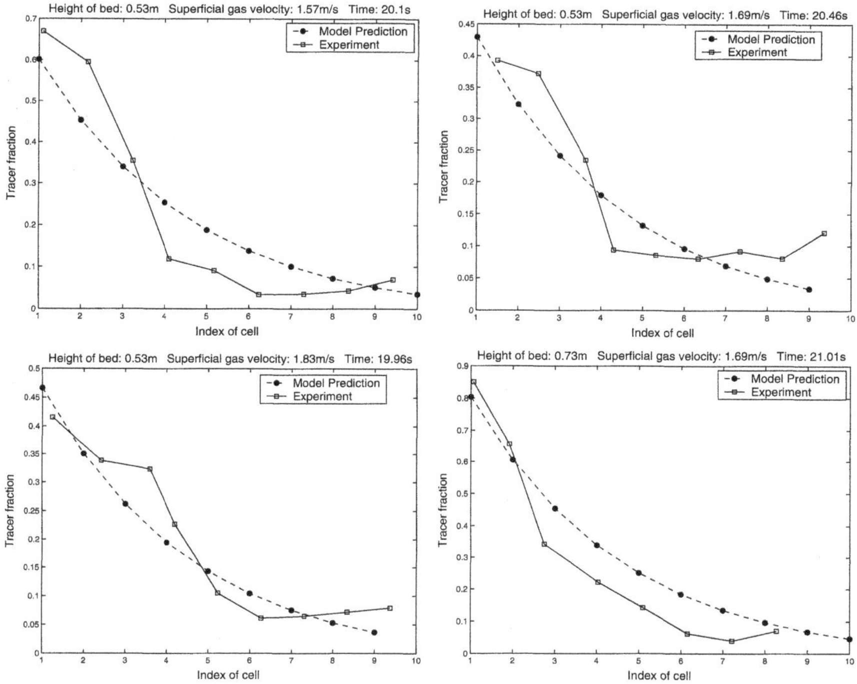

Figure 3.12: Distribution of jetsam volume fraction with $\alpha = 0.17$ and $r = 1$.

part of the bed increases slightly with bed height when in the model predictions the tracer fraction is a decreasing function of bed height. It should be possible to achieve even better predictions with more knowledge on how to choose r and α.

Let us focus briefly on the two adjustable parameters r and α. Figure 3.13 shows the sensitivity of the model predictions to their values. To the left the parameter r, which describes the rate of flow into annulus region, is varied. The effect of r is seen to be somewhat limited for this case. The figure to the right shows the more significant model sensitivity to variation in the parameter α, which determines the rate of segregation. There is no segregation for zero α and segregation increases with α. Overall, Figure 3.13 shows that the parameters play reasonable roles in the model, the predictions are neither overly sensitive nor overly insensitive to the value of parameters.

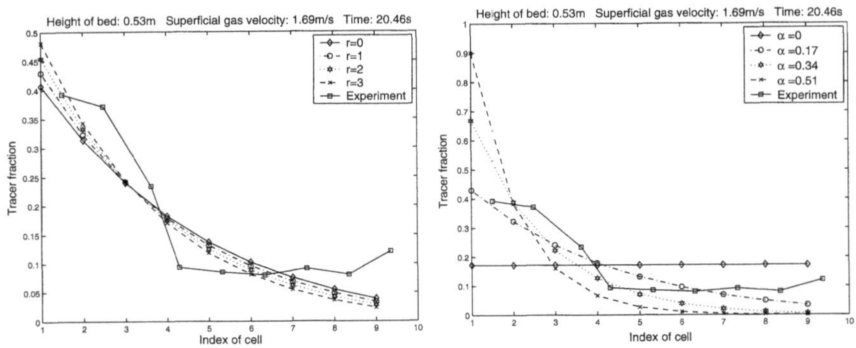

Figure 3.13: Distribution of jetsam volume fraction for different α and r.

To validate the stochastic model further, it was compared with extra series of experimental data from a completely different system. The system is a slugging bed containing coal and limestone [2]. In this system coal acts as flotsam and limestone as jetsam. Figure 3.14 shows comparisons of the experimental data with model predictions. The experiments were carried out in a column of 13 cm diameter and were performed at two superficial gas velocities, 0.8 and 0.95 m/s, and the initial bed height of 0.75 m. Good agreement is seen also here.

Even though the experimental results and model prediction agree well, discrepancies can be seen in the bottom of the bed in both systems. In both systems an increasing fluidization velocity affects the concentration of particles at the bottom of the bed the most and affects it only slightly in the higher part of the bed. The model does not reflect this. This can be explained since the model has not taken the bubble-forming-a-slug region into account. We assume that the slug as a whole forms from above the distributor while, as mentioned before, in the actual system it is likely that a region of bubbling fluidization exists below the slug forming region. The slugs are formed from coalescence of small bubbles and this process is affected by a change in the fluidization velocity.

Regarding the segregation effect in Figure 3.14, the coal-limestone system, we used $\alpha = 0.07$ instead of $\alpha = 0.17$. This is reasonable, since this system is completely different from the other one, a difference in segregation can be expected. In the model a lower value of α reflects a stronger segregation. This agrees qualitatively with experiment, as we now show by means of the segregation distance for bubbling beds.

From Equation (3.9) for the segregation distance $\bar{Y}_s$ we find that at the

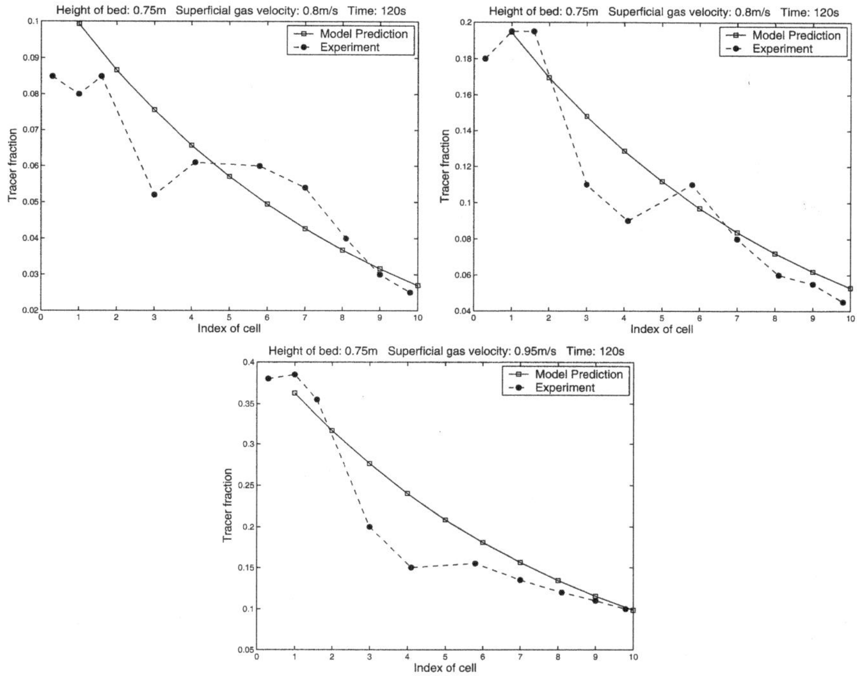

Figure 3.14: Distribution of jetsam volume fraction for $\alpha = 0.07$ and $r = 1$

same jetsam concentration, $x_0 = 0.2$, $\bar{Y}_s^{pa} \approx 0.24$ for the pigment agglomerate system while the coal-limestone system has a segregation distance $\bar{Y}_s^{cl} \approx 0.74$. Even though our system is not a bubbling bed this confirms that the pigment agglomerate system should exhibit less segregation than the coal-limestone system.

Figures 3.12 and 3.14 represent different types of particles and bed geometries. Both agree very well with the model prediction apart from the velocity effect in the lowest part of the bed mentioned above. This confirms that the stochastic model is suitable for slugging fluidized beds in general.

Turning now to the issue of predicting the of r and α quantitatively. Little information for quantifying r can be found in the literature at present. r is not likely to change much with the size of the slug, but the distance between successive slugs may have an effect. As far as α is concerned, we have mentioned above that the distance (reduced with the bubble radius) that a particle segregates due to a single fluidization bubble in a bubbling bed varies with: $(d_{bulk}/d_{particle})^{1/3}(\rho_{bulk}/\rho_{particle})$. In a slugging bed, we would not expect that the absolute segregation distance would change much with the height of the slug, but its variation with the particle properties may be similar to that quoted for a bubbling bed.

When comparing the transient behaviour of the model with experiment we find that the model profiles seem to reach a steady state faster than the experimental profiles. This could possibly depend on the choices of α and r.

We note that one assumption of this model is that the jetsam is infinitely dilute so that:

1. All the jetsam particles can collect in one cell (i.e. no interaction between particles in the sense that they can fill a cell).

2. The segregation rate is not influenced by the presence of other jetsam particles (the local environment is not influenced by the concentration of jetsam particles).

The issue of interfering particles is discussed in Section 3.4.

In summary, the stochastic approach has been used to formulate a model for a system so complex that modeling with the deterministic approach would have been difficult if not impossible. It has a further important attraction in significantly reduced calculation time when compared to the alternatives. These advantages are present both for systems that exhibit significant stochastic behaviour and for systems that, through the law of large numbers, exhibit essentially deterministic behaviour in practice.

3.4 Stochastic Model Incorporating Interfering Particles

In this section we present on-going work on an interacting particle model for transport of particles in a binary fluidized bed reactor. The model incorporates physical processes that were also the basis of the stochastic models for single particle transport presented in Chapters 3 and 5. We apply the model to investigate particle separation in binary fluidized beds, and compare model predictions with experimental results.

3.4.1 Introduction

In a series of papers [34, 35, 39, 62], transport phenomena in bubbling fluidized bed reactors have been studied with the help of models for the motion of a single particle. This work has been described in Chapters 3 and 5 in this book. These models have also been used to predict the diffusion of a pulse of marked particles, assuming independence of the particles in motion. This is justified so long as the proportion of marked particles is rather small, as was the case in all the experiments analysed and described in Hoffmann and Dehling [62] and Dechsiri et al. [36–38, 61]. There the single particle model was very successful in predicting the nature of the results of the experiments.

When there is a sizable fraction of marked particles in the reactor, however, interaction among particles becomes relevant. The simple fact that no two particles can occupy the same location implies that there is an upper bound on the local proportion of marked particles. In Section 3.2, the problem of interaction was addressed in a rather ad hoc manner. The goal of the present paper is to study a particle transport model that incorporates interaction effects in a fundamental way.

3.4.2 Physical transport processes in fluidized bed reactors

The model is based on the particle transport processes discussed in Section 3.2.1, such that all the transport mechanisms are associated with the action of fluidization bubbles rising through the bed. This is transport up in the bubble wake phase, transport down in the bulk, and dispersion due to the stirring action of the rising bubbles. When the bed, as here, consists of particles with differing physical properties, there is superimposed on the other transport mechanisms extra motion due to the particles' relative movement: downward movement of the denser and/or larger particles tending to sink and upward motion of the lighter and/or finer particles. This latter motion is also only possible in the regions disturbed by fluidization bubbles rising in the vicinity (see Hartholt et al. [58]).

3.4.3 Particle exchange model for diffusion

To present the essential ideas, we first investigate a model for only diffusional transport, and do not take the transport in the wake phase into account. Incorporating wake transport will be done in one of the following sections. Moreover we concentrate on a batch fluidized bed which for modeling we divide into N horizontal layers (cells) of equal height. We assume that there are two types of particles in the reactor, say marked and unmarked particles (Figure 3.15).

We further assume that the marked particles due to their larger density tend to segregate toward the bottom of the bed. For simplicity we assume that each cell holds exactly K particles. In total the reactor contains $N \cdot K$

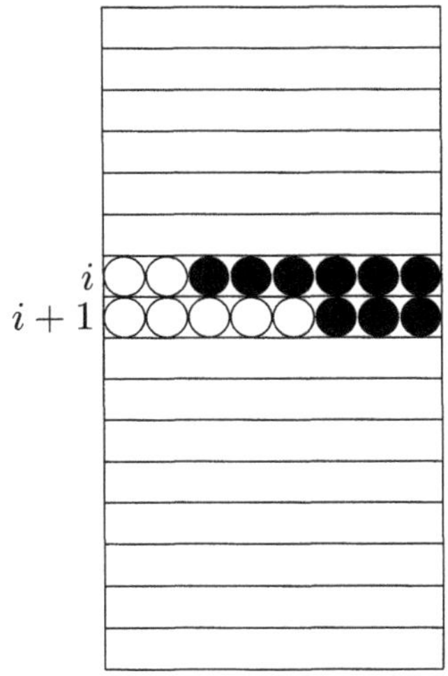

Figure 3.15: Particle exchange model for diffusion with discretized reactor

particles, and we assume that M of these are marked. We denote by $Y_t(i)$ the number of marked particles in cell i at time t. The state of our system at time t is thus described by the vector $Y_t := (Y_t(1), ..., Y_t(N))$.

The basic transport mechanism in our model is via the exchange of particles between neighboring cells. In this exchange, marked particles are preferentially chosen for downward transport and this will result in segregation of particles. Particle exchange starts by choosing a cell at random, its number denoted by i, where $i \in 1, ..., N-1$. We then chose one particle from cell i and one from cell $i+1$ using a weighted probability distribution, and have these particles exchange positions. The particle selected from cell i is a marked particle with probability

$$\frac{\alpha\, y(i)}{\alpha\, y(i) + K - y(i)} = \frac{\alpha\, y(i)/K}{\alpha\, y(i)/K + 1 - y(i)/K} = \frac{\alpha\, c(i)}{\alpha\, c(i) + 1 - c(i)}$$

and unmarked with probability

$$\frac{K - y(i)}{\alpha\, y(i) + K - x(i)} = \frac{1 - c(i)}{\alpha\, c(i) + 1 - c(i)}.$$

In these formulae, $c(i) := \frac{y(i)}{K}$ denotes the concentration of marked particles in cell i and $\alpha \geq 1$ is a constant governing the degree of preference for a marked particle to be chosen for downward transport. Observe that $\alpha = 1$ would imply that the particle is selected completely at random from the K particles in cell i.

Similarly, we assume that the particle chosen from cell $i+1$ for upward transport is marked with probability

$$\frac{\beta\, c(i+1)}{\beta\, c(i+1)+1-c(i+1)},$$

and unmarked with probability $\frac{1-c(i+1)}{\beta\, c(i+1)+1-c(i+1)}$, where $\beta \leq 1$ is again a weight factor.

The above defines a Markov process $\underline{Y}_t = (Y_t(1), ..., Y_t(N))$ with parameter space $\mathbb{N}_0$ and state space $S = \{(y(1), ..., y(N)) : y(i) \in \mathbb{N}_0 : \sum_{i=1}^{N} y(i) = M\}$. This state space is too big to allow numerical calculations of particle densities as in the earlier work listed and thus we have pursued a simulation study.

In figure 3.16 we show the marked particle concentration after 1,000,000 iterations for a model with $N = 20$, $K = 400$, $M = 4.000$, $\alpha = 1.2$, $\beta = .8$ and a uniform initial distribution of marked particles in the NK positions. Convergence for a larger number of cells is rather slow and more attention should be paid to find efficient algorithms.

3.4.4 Invariant density for a large number of particles

As the number of transitions increases, the distribution of the Markov process will converge to a steady state. This steady state distribution fluctuates around the infinite particle limit, and fluctuations decrease at $\frac{1}{\sqrt{K}}$ rate as we let the number of particles go to infinity, i.e. $N \to \infty$, $K \to \infty$, $K/N \to \gamma \in [0, 1]$.

In what follows we will compute this infinite particle limit.

Still denoting the concentration of marked particles in cell i by $c(i)$, we get in our model a transition from $y(i)$ to $y(i)+1$ if a marked particle from cell $i+1$ exchanges position with an unmarked particle from cell $i+1$. This event has probability

$$\frac{\beta\, c(i+1)}{\beta\, c(i+1)+1-c(i+1)}\, \frac{1-c(i)}{\alpha\, c(i)+1-c(i)} \tag{3.36}$$

In a similar way, we get that a transition from $y(i)$ to $y(i)-1$ occurs with probability

$$\frac{\alpha\, c(i)}{\alpha\, c(i)+1-c(i)}\, \frac{1-c(i+1)}{\beta\, c(i+1)+1-c(i+1)} \tag{3.37}$$

The event of no change in the distribution of marked particles occurs when either two marked or two unmarked particles exchange positions, and this

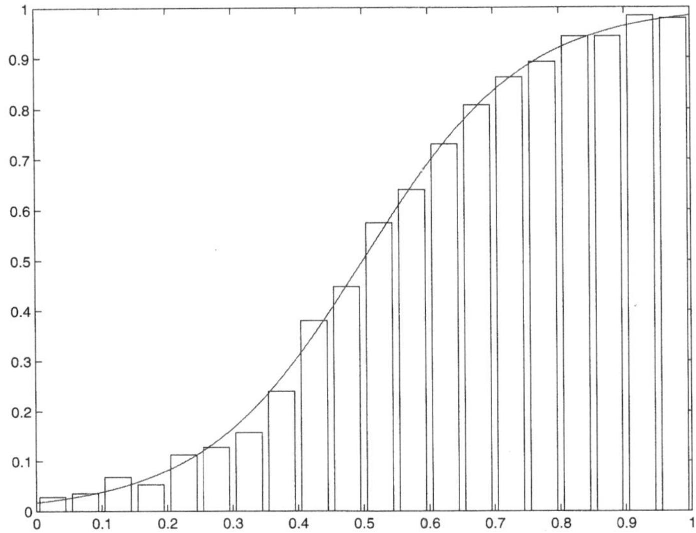

Figure 3.16: Density of marked particles in a pure diffusion model with segregation from 1,000,000 simulations for a model with $N = 20$, $K = 400$, $M = 4.000$, $\alpha = 1.2$, $\beta = .8$ and a uniform initial distribution of marked particles. The smooth curve is that found in the previous section for this example.

has probability

$$\frac{\alpha\, c(i)\, \beta\, c(i+1) + (1-c(i))\,(1-c(i+1))}{(\alpha\, c(i) + 1 - c(i))\,(\beta\, c(i+1) + 1 - c(i+1))}$$

A steady state is reached if there is no drift, and this means that (3.36) and (3.37) must be equal. We thus arrive at the following condition,

$$\alpha\, c(i)\,(1-c(i+1)) = \beta\, c(i+1)\,(1-c(i))$$

which can be rewritten as

$$\frac{c(i+1)}{1-c(i+1)} = \frac{\alpha}{\beta}\,\frac{c(i)}{1-c(i)}.$$

This iteration can be solved explicitly by

$$\frac{c(i)}{1-c(i)} = (\frac{\alpha}{\beta})^i\,\frac{c(0)}{1-c(0)} = e^{\gamma_0+\gamma_1\, i}, \tag{3.38}$$

where $\gamma_0 = \log \frac{c(0)}{1-c(0)}$ and $\gamma_1 = \log \frac{\alpha}{\beta}$. We can solve (3.38) explicitly for c(i) and get

$$c(i) = \frac{e^{\gamma_0 + \gamma_1 i}}{1 + e^{\gamma_0 + \gamma_1 i}}.$$

If we consider the continuous limit obtained by letting the space discretization converge to zero, or equivalently $N \to \infty$, the concentration of marked particles in the steady state becomes

$$c(x) = \frac{e^{\gamma_0 + \gamma_1 x}}{1 + e^{\gamma_0 + \gamma_1 x}}, \; 0 \leq x \leq h, \tag{3.39}$$

where $\gamma \in \mathbb{R}$ and $\gamma_1 \geq 0$ are parameters.

If the average concentration of marked particles in the reactor is known, this puts a restriction on the parameters γ_0 and γ_1. Suppose the average concentration is C, then we get

$$\begin{aligned} C = \frac{1}{h}\int_0^h c(x)dx &= \frac{1}{h}\int_0^h \frac{e^{\gamma_0+\gamma_1 x}}{1+e^{\gamma_0+\gamma_1 x}}dx \\ &= \frac{1}{h\gamma_1}\log(1+e^{\gamma_0+\gamma_1 x})\Big|_{x=0}^{x=h} \\ &= \frac{1}{h\gamma_1}\log\frac{1+e^{\gamma_0+\gamma_1 h}}{1+e^{\gamma_0}}. \end{aligned}$$

We can solve this to get an explicit equation for γ_0. Via some calculations we easily find

$$e^{h\gamma_1 C}(1 + e^{\gamma_0}) = 1 + e^{\gamma_0}\, e^{\gamma_1 h}$$

and thus $e^{\gamma_0} = \frac{e^{h\gamma_1 C} - 1}{e^{\gamma_1 h} - e^{\gamma_1 hC}}$, from where we finally obtain

$$\gamma_0 = \log\left(\frac{e^{h\gamma_1 C} - 1}{e^{\gamma_1 h} - e^{\gamma_1 hC}}\right).$$

In figure 3.17 we have drawn graphs of the concentrations for the fixed average concentration of $C = 50\%$ and a variety of values of the parameter γ_1.

3.4.5 Comparison with data

If we want to fit this model to data, we can take logarithms on both sides of (3.38) and obtain

$$\log\left(\frac{c(i)}{1-c(i)}\right) = \log\frac{c(0)}{1-c(0)} + i\,\log\left(\frac{\alpha}{\beta}\right) = \gamma_0 + \gamma_1\, i.$$

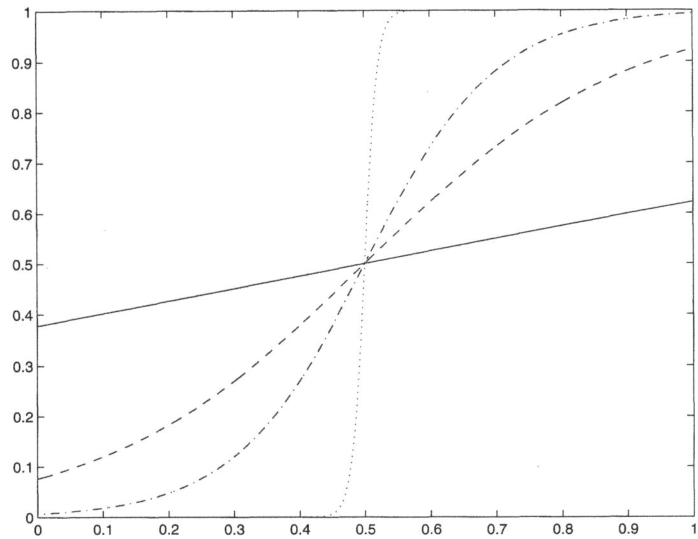

Figure 3.17: Invariant densities according to the pure diffusion model for average concentration $C = 0.5$ and $\gamma_1 = 1$(solid line), $\gamma_1 = 5$ (dashed line), $\gamma_1 = 10$ (dash-dotted line) and $\gamma_1 = 100$ (dotted line)

It is therefore possible to determine γ_0 and γ_1 by a least squares linear regression of $\log\left(\frac{c(i)}{1-c(i)}\right)$ on i.

Comparing the data (see figure 3.18) to our model prediction we find that the near constant concentration of marked particles at the top of the reactor cannot be well fit by our model. We can interpret the fact that particles at the top of the reactor are almost perfectly mixed as a result of wake flow that will continuously transport particles from the bottom of the reactor to the top.

3.4.6 Particle model for diffusion and wake flow

In this section we add wake flow to the model discussed in the previous section. For the sake of simplicity, we assume that only particles in the bottom cell can move upward in the wake flow. They are then deposited at the top of the reactor. As a result there is a downward flow of particles to fill the void left by the wake flow.

To put the wake flow and diffusion effects on the same scale, we set the

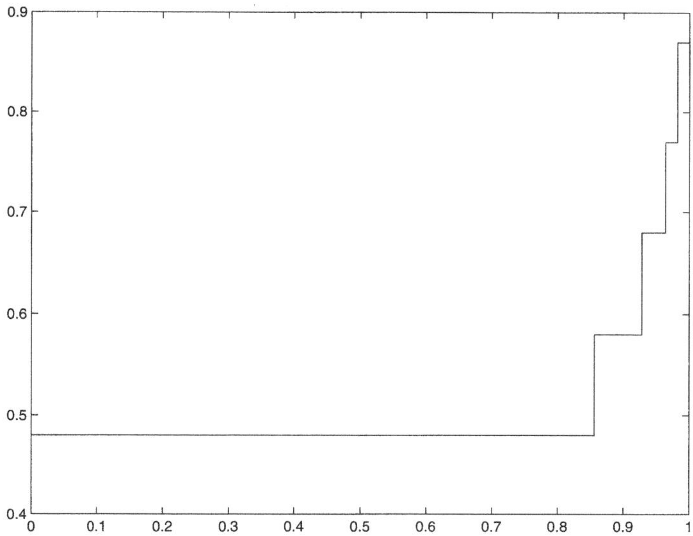

Figure 3.18: Concentration of jetsam particles in a fluidized bed reactor, experiment by C. Dechsiri

basic time unit to N cycles of the exchange of particles between adjacent cells. Then on average every pair of cells has been chosen once during the time unit. After each such cycle, a Poisson number of particles is picked from the bottom cell, here each particle is with probability $\frac{y(N)}{K}$ marked and with probability $1 - \frac{y(N)}{K}$ unmarked. Denoting these numbers by U and V respectively and $W := U + V$, we have the following distributional properties:

$$\begin{aligned} W &\sim Poisson(\lambda) \\ U &\sim Poisson(\lambda \frac{y(N)}{K}) \\ V &\sim Poisson(\lambda (1 - \frac{y(N)}{K})) \end{aligned}$$

and U,V are two independent random variables. These particles are then deposited at the top and the void left at the bottom induces a downward flow of particles. We model this downward flow as in section 3.4.1. Each of the W particles that move from cell i to cell $i+1$ is marked with probability $\frac{a\,y(i)}{a\,y(i)+K-y(i)}$ and unmarked with probability $\frac{K-y(i)}{a\,y(i)+K-y(i)}$. When applying this, one has to realize that the choice of particles that move influences the possible moves in neighbouring cells so that the order is important. We start from $i = N - 1$ and go upward to $i = 1$.

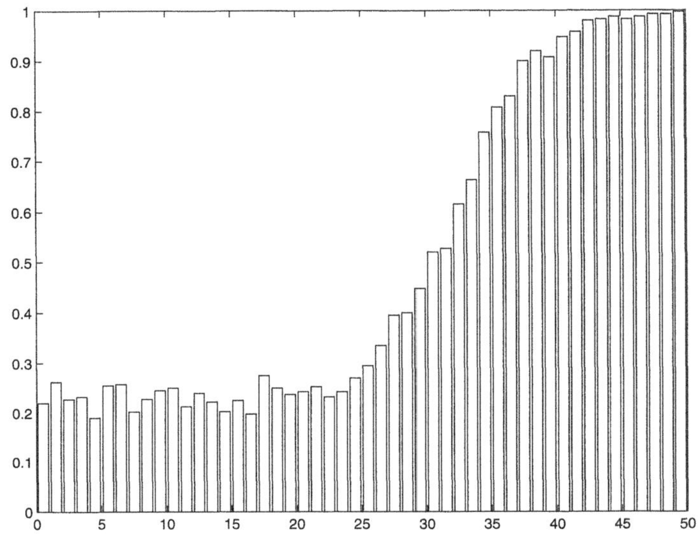

Figure 3.19: Density of marked particles in the model including wake flow, simulation of the model with $N = 50$, $K = 400$,$M = 10,000$,$\alpha = 1.2$, $\beta = 0.8$, $\lambda = 0.1$,$S = 100,000$ simulations.

Chapter 4

Continuous Systems and Residence Time Distribution

In this and the following chapter we will consider residence time distribution (RTD) in processing vessels with continuous throughflow. P.V. Danckwerts laid much of the groundwork for this field of study in his paper from 1953[30]. After introducing his work, we discuss subsequent investigations, emphasizing stochastic aspects, and elucidate Danckwerts' law for the mean residence time using a stochastic approach. In the following chapter we will give an account of a stochastic model for particle residence time in continuous fluidized beds.

4.1 Theory of Danckwerts

We start this section with a presentation of Danckwerts' basic results. We then discuss his stochastic model for blenders in a separate subsection.

4.1.1 Danckwerts' Basic Results

To derive expressions for the RTD of fluid elements in a processing vessel with *continuous and steady* throughflow, Danckwerts considered the situation where the inflow, which is volumetrically constant at Q m^3/s, to the vessel of volume V undergoes a step change in composition, say from all white to all red fluid. He called a plot of the fraction $F(t)$ of red material in the outflow an "F-diagram". F will obviously approach unity for long times.

For perfect plug flow (or "piston flow", flow with a cross-sectionally uniform velocity) in a tube, F will undergo a step change from 0 to 1 at

$t = V/Q$. We define a dimensionless time: $\theta := t(Q/V)$[1]. For some degree of mixing between in- and outlet, the F curve will be smooth. See the sketches in Figure 4.1.

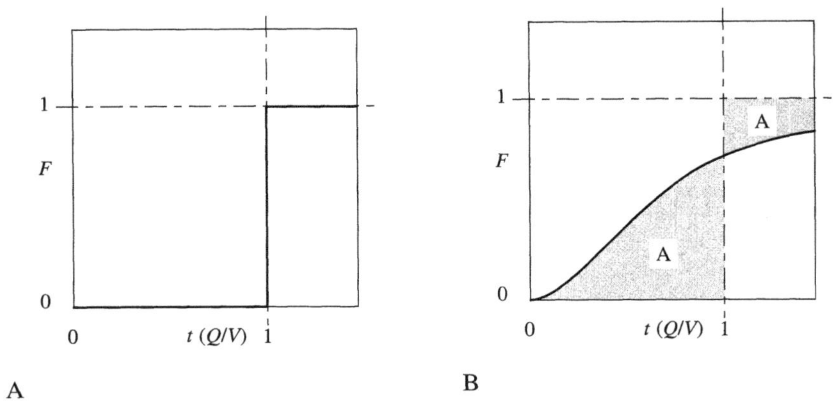

Figure 4.1: Sketches of F-curves for **A**: piston flow through tube and **B**: a degree of mixing between inlet and outlet

An important aspect in Danckwerts' analysis is that the two shaded areas in the figure, under the F-curve between $0 \leq \theta < 1$ and between the F-curve and 1 at $\theta > 1$, respectively, are equal. That this is so can be seen in the following way.

Since F is the fraction of red material in the outgoing stream, the flow of red material during a time element dt is $QFdt$. The total volume of red material flowing out during $0 \leq t < V/Q$ is then $\int_0^{V/Q} QFdt$. The total volume of white material flowing out during this period is

$$\int_0^{\frac{V}{Q}} Q(1-F)dt = V - \int_0^{\frac{V}{Q}} QFdt. \tag{4.1}$$

If there are no completely dead zones in the vessel, the rest of the white material in the vessel must flow out in the period $t \geq V/Q$, and thus:

$$\int_{\frac{V}{Q}}^{\infty} Q(1-F)dt = V - \left(V - \int_0^{\frac{V}{Q}} QFdt\right) = \int_0^{\frac{V}{Q}} QFdt. \tag{4.2}$$

[1]Note that Danckwerts denotes time itself by θ and uses no separate symbol for dimensionless time

Changing the variable of integration from t to θ and simplifying gives the desired result:

$$\int_1^\infty (1-F)d\theta = \int_0^1 F d\theta. \tag{4.3}$$

In addition to $F(t)$, Danckwerts also defines two other functions, the "exit age" and "internal age" distributions, $E(t)$ and $I(t)$. These are density functions, such that the fractions of fluid elements in the exit stream and in the internal of the vessel, respectively, with residence times ("ages") between t and $t+dt$ are $E(t)dt$ and $I(t)dt$.

According to its definition, $F(t)$ can be seen as the fraction of fluid elements in the exit stream with ages less than t, so that:

$$F(t) = \int_0^t E(s)ds. \tag{4.4}$$

E-curves are sketched in Figure 4.2. The E-curve for piston flow is a Dirac delta function, δ, which has the properties:

$$\begin{aligned}
&\delta(x-a) = 0 && \text{for } x \neq a \\
&\delta(x-a) = \infty && \text{for } x = a \\
&\int_{a-\varepsilon}^{a+\varepsilon} f(x)\delta(x-a)dx = f(a) && \text{for } \varepsilon > 0,
\end{aligned}$$

To relate $I(t)$ to the other functions, a balance for red material, *i.e.* material with age less than t, is performed:

$$\text{total in} = \text{still in vessel} + \text{total out}$$

$$Qt = V\int_0^t I(s)ds + Q\int_0^t F(s)ds. \tag{4.5}$$

Differentiating w.r.t. t, dividing by Q and rearranging gives the desired relation:

$$1 - F(t) = \frac{V}{Q}I(t). \tag{4.6}$$

The *mean residence time* of fluid elements is:

$$\langle t\rangle = \int_0^\infty tE(t)dt = \int_0^\infty t\frac{dF(t)}{dt}dt. \tag{4.7}$$

Writing t in terms of θ, changing the variable of integration to θ and dividing through by V/Q gives:

$$\langle\theta\rangle = \int_0^\infty \theta\frac{dF(\theta)}{d\theta}d\theta = \int_0^1 \theta dF(\theta) = 1. \tag{4.8}$$

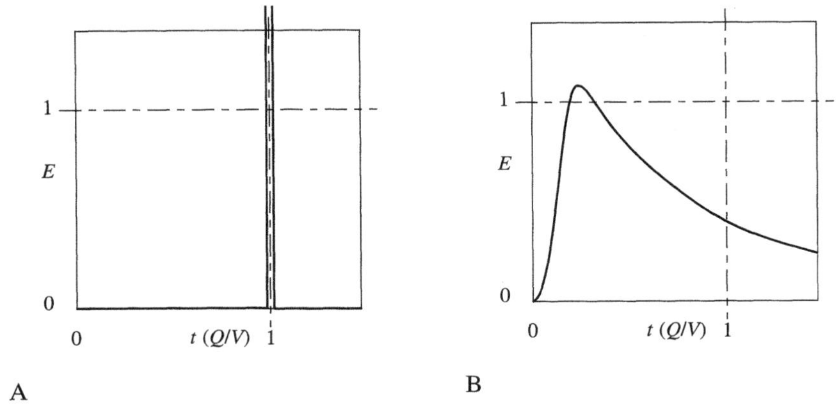

Figure 4.2: Sketches of E-curves for **A**: piston flow through tube, the E-curve is a Dirac delta function and **B**: a degree of mixing between inlet and outlet

The last result is evident from Figure 4.1.

This ends our brief discussion of Danckwerts' basic results. Danckwerts continues to define two parameters to characterize the performance of a reactor. These parameters are not our direct interest here. The rest of Danckwerts' paper is dedicated to special cases, which we discuss now.

One of these special cases is an *ideally mixed* vessel, also called a "Continuous Stirred Tank Reactor" or CSTR (see also *e.g.* Westerterp et al [135] or Levenspiel [83]). A simple material balance over such a vessel shows that the F-curve is: $F(t) = 1 - \exp(-Qt/V)$. We discuss this later in this chapter and in Appendix A.

Another special case is the derivation of an expression for $F(t)$ for *laminar flow through a tube section* simply from the laminar velocity profile, which can be found analytically [17]. This does not fall under the heading of "stochastic models" and will not be discussed here.

The case of *flow through packed particle beds* is by Danckwerts treated as *plug flow with dispersion* (also called "Dispersed Plug Flow" or DPF), described by Equation (1.10). Danckwerts solves this case by appealing to "penetration theory" for diffusion into a semi-infinite medium. Discussions of penetration theory can be found in Beek et al. [6] or Bird et al. [17], and we give a solution method for DPF, which is very similar to that of Danckwerts in Appendix A.

For the case of *flow through packed beds with a first order reaction*, Danckwerts no longer calculates F- or E-curves, but rather the steady state concentration of reactant remaining at the outlet. To do this he solves Equation (1.10) for steady flow, but with an extra term accounting for the disappearence of reactant due to reaction:

$$0 = -v_x \frac{\partial C(x,t)}{\partial x} + D \frac{\partial^2 C(x,t)}{\partial x^2} - kC(x,t), \tag{4.9}$$

where k is a reaction rate constant. We will not discuss this case, but note that it illustrates the fact that if a first order chemical reaction takes place in the vessel or reactor under consideration, the concentration of reactant at the outlet—and therefore the degree of conversion in the reactor—can be calculated directly from the RTD.

4.1.2 Danckwerts' Stochastic Model for Blenders

A blender is often used in the process industry to suppress variations in product quality in the outgoing stream. For instance variability in the quality of natural gas can be suppressed by inserting into the product stream a large vessel to act as a blender. Mixing takes place in the blender so that the properties of the outgoing stream becomes more uniform in time.

To analyse the action of a blender, Danckwerts divides the concentration of a component, C, in the in- and outgoing streams into a mean part and a fluctuating part, the mean of the fluctuating part being zero; for instance for the ingoing stream: $C_{in} = \langle C \rangle_{in} + \tilde{C}_{in}$. Let σ^2 be the variance of $\tilde{C}_{in}$. The concentration in the outgoing stream is:

$$\begin{aligned} C_{out}(t) = \langle C \rangle_{out}(t) + \tilde{C}_{out}(t) &= \int_0^\infty C_{in}(t-s)E(s)ds \\ &= \int_0^\infty \left(\langle C \rangle_{in}(t-s) + \tilde{C}_{in}(t-s) \right) E(s)ds \end{aligned} \tag{4.10}$$

If $\langle C \rangle$ is invariant in time, $\langle C \rangle_{in} = \langle C \rangle_{out} = \langle C \rangle$ and the terms involving it will cancel. Squaring the rest gives:

$$\begin{aligned} \left(\tilde{C}_{out}(t) \right)^2 &= \left(\int_0^\infty \tilde{C}_{in}(t-s)E(s)ds \right)^2 \\ &= 2 \int_0^\infty \int_0^\infty \tilde{C}_{in}(t-s)E(s)\tilde{C}_{in}(t-(s+r))E(s+r)dsdr. \end{aligned} \tag{4.11}$$

The last result is from basic calculus for the autocorrelation function. The inner integral on the right-hand-side is the autocorrelation function of $\tilde{C}_{in}(t-s)E(s)$ on the interval $[0,\infty)$, which is an even function of r.

The time-mean of $\left(\tilde{C}_{out}(t)\right)^2$ is simply the variance of the fluctuating concentration in the outgoing stream, σ_{out}^2. The "autocorrelation coefficient" for the function $\tilde{C}_{in}(t)$ is defined as:

$$R(r) := \frac{\langle \tilde{C}_{in}(t-s)\tilde{C}_{in}(t-(s+r))\rangle}{\sigma_{in}^2}. \tag{4.12}$$

Thus averaging with respect to time finally gives:

$$\left(\frac{\sigma_{out}}{\sigma_{in}}\right)^2 = 2\int_0^\infty \int_0^\infty R(r)E(s)E(s+r)dsdr \tag{4.13}$$

The autocorrelation coefficient normally has the form sketched in Figure 4.3. It makes intuitive sense that the fluctations in the outlet stream should be larger for larger time-scale fluctations in the inlet stream.

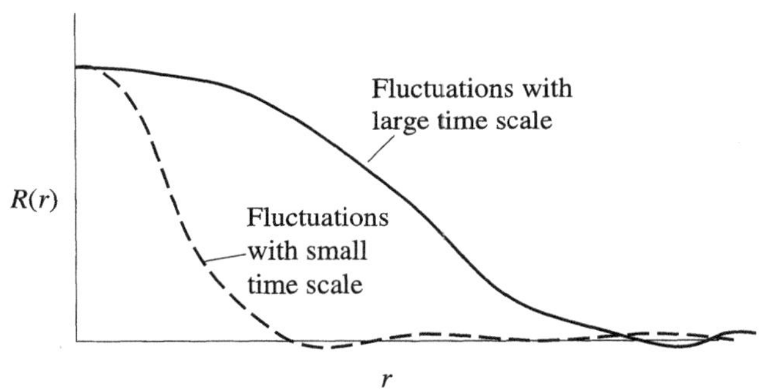

Figure 4.3: Sketches of autocorrelationcoefficients with large and small time-scale fluctations.

4.2 Subsequent Work

During the subsequent period, RTD analysis became an important tool in process diagnostics. Two models, which some reviewers classify as "stochastic RTD models" became especially important:

- ideal mixers in series, also called "continuous stirred tank reactors", CSTR, and

- plug flow (in a tube section) with axial dispersion, also called "dispersed plug flow", DPF.

4.2.1 CSTRs and DPF

The reason for the popularity of these two models is that they both can be fitted to a wide range of processes by varying just one dimensionless parameter.

We give accounts of both in Appendix A. The equations derived there for F are:

$$F = 1 - e^{-n\theta} \sum_{i=1}^{n} \frac{(n\theta)^{i-1}}{(i-1)!}, \tag{4.14}$$

for n CSTRs in series, and:

$$F = -\frac{1}{2}\text{erf}\left(\frac{1-\theta}{\sqrt{\frac{4}{Pe}\theta}}\right) + \frac{1}{2}, \tag{4.15}$$

for the DPF. The parameters n and Pe are normally fitted to experimental residence time distributions to create semi-empirical RTD models for a given process.

In accordance with Equation (4.4) we can calculate E-curves from the above formulae by differentiating w.r.t. θ. Figure 4.4 shows E- and F-curves using parameter values for which the shapes of the curves are similar for the two models. Clearly there is not much to choose between them when fitting to experimental data. The E-curves from the DPF-model are slightly more positively skewed than the ones from the CSTR-model.

An important distinction in RTD theory is between systems with "closed" and "open" boundaries, respectively. In the former, a particle or fluid element cannot reenter the system through an outflow boundary, and not exit through an inflow boundary, while in the latter it can. A CSTR has closed boundaries, while the outflow boundary in the DPF is open: the flow over the "outflow" boundary is really a *net* outflow, the total outflow due to convection and diffusion together minus the inflow through diffusion.

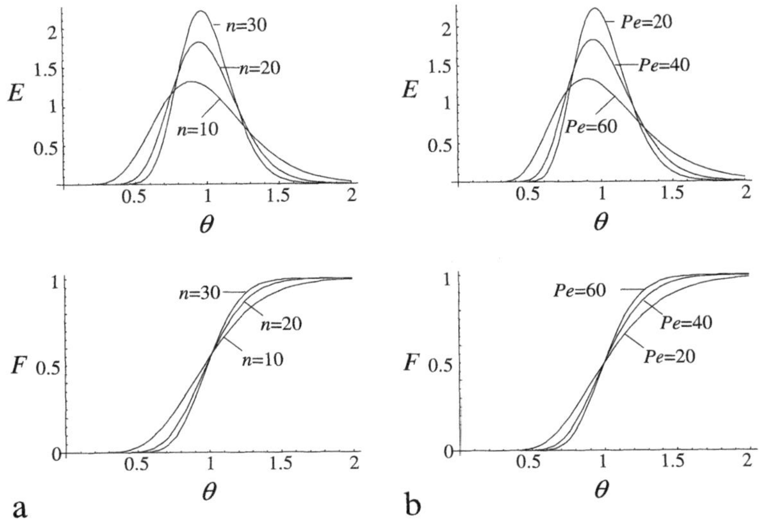

Figure 4.4: Calculated E- and F-curves for **a**: the CSTR-model with different values for n **b**: the DPF-model with different values for Pe

4.2.2 Review of Torres and Oliveira

Material RTD is especially important in the food industry. Torres and Oliveira[121] have written a comprehensive review of both experimental and theoretical investigations of RTD in this sector of the processing industry. They classify models in:

- "Mechanistic models", these are RTD models derived by calculation from the flow field in the vessel, an example is Danckwerts' above-mentioned derivation of the RTD for laminar flow through a tube on basis of the fluid velocity profile.

- "Stochastic models", RTD models derived from a stochastic perspective, even if the deterministic limit has been taken. The two types of models mentioned above, CSTRs in series and DPF are classified as stochastic models.

- "Probabilistic models", these are models where the experimentally determined F-curve has simply been fitted by some probabilistic distribution, such as the normal or the Poisson distributions.

4.2.3 Gibilaro's Extension to Danckwerts' Mean Residence Time

Gibilaro[47, 48] extended Danckwerts' law for mean residence time expressed in Equation (4.8). He found that in a vessel with closed boundaries, the average residence time of the fluid in any subregion of the vessel is:

$$\langle t \rangle = \frac{V_s}{Q}, \tag{4.16}$$

where V_s is the volume of the subregion, and Q, as before, is the total volumetric flow through the vessel.

He showed this by considering a first order reaction of a component A:

$$\mathrm{A} \xrightarrow{k} \mathrm{B},$$

taking place only in the subregion of interest, one might imagine some catalyst necessary for the reaction to be present only in the region of interest, evenly distributed over the region but not taking up any of its volume so that the whole region is available for flow.

Thus for the concentration of A, C_A in a fluid element residing in that region, we can write:

$$-\frac{dC_A}{dt} = kC_A \Rightarrow -\ln C_A = kt \Rightarrow C_A = e^{-kt}. \tag{4.17}$$

We call the exit age distribution in the region $E_s(t)$, so that $E_s(t)dt$ is the material fraction in the stream exiting the vessel that has resided in the subregion between t and $t + dt$. Then the mean concentration of A in the stream leaving the vessel is:

$$C_{A,o} = C_{A,i} \int_0^\infty E_s(t) e^{-kt} dt, \tag{4.18}$$

where $C_{A,i}$ is the constant concentration of A in the material entering. Differentiating w.r.t. k gives:

$$\frac{dC_{A,o}}{dk} = -C_{A,i} \int_0^\infty tE_s(t) e^{-kt} dt, \tag{4.19}$$

and letting k go to zero gives an expression for the mean residence time in the subregion:

$$-\frac{1}{C_{A,i}} \frac{dC_{A,o}}{dk}\bigg|_{k \to 0} = \int_0^\infty tE_s(t)dt = \langle t \rangle. \tag{4.20}$$

Performing a mass balance over the entire vessel yields:

$$\text{inflow} = \text{outflow} + \text{dissappearance through reaction}$$
$$QC_{A,i} = QC_{A,o} + k\int_{V_s} C_A dv. \tag{4.21}$$

Differentiating this with respect to k gives:

$$\frac{dC_{A,o}}{dk} = -\frac{1}{Q}\left(\int_{V_s} C_A dv + k\frac{d}{dk}\int_{V_s} C_A dv\right), \tag{4.22}$$

since $C_{A,i}$ does not depend on k. Letting k go to zero, C_A tends to $C_{A,i}$, and its dependence on k vanishes leaving:

$$-\frac{1}{C_{A,i}}\frac{dC_{A,o}}{dk}\bigg|_{k\to 0} = \frac{\int_{V_s} dv}{Q} = \frac{V_s}{Q}. \tag{4.23}$$

Comparing Equations (4.20) and (4.23) gives the desired result, Equation (4.16).

Gibilaro also introduces the notion of *stochastically independent subregions* as subregions for which the combined exit age distribution can be found from the individual exit age distributions by convolution:

$$E(t) = E_{s_1}(t) * E_{s_1}(t) * \ldots * E_{s_n}(t), \tag{4.24}$$

meaning that the probability of a fluid element's residence time in one subregion is independent of its residence time in another.

4.2.4 Fan et al.'s Stochastic Model for Transient RTD

Fan et al. [43] performed a stochastic analysis of transient RTD in processing vessels. They point out that RTD theory, although in nature a stochastic process, is normally formulated using deterministic terminology and deterministic concepts.

If the flowing medium is viewed as consisting of *discrete particles* (fluid elements or molecules, for instance), then the "life" (residence time) of a particle in the vessel can be modeled by a stochastic variable, called Θ by Fan et al. with state-space $[0, \infty)$. The function $F(t)$ is then the cumulative probability function for this variable: $F_\Theta(t) := P(\Theta \le t)$, or the expected value for the fraction of particles with a residence time less than t. Danckwerts' exit age distribution, $E(t)$, is the probability density function for Θ, called $f_\Theta(t)$ by Fan et al., a notation we will adopt in this section.

The main point of their work was to consider "start-up effects" in a reactor that is already full, and at $t = 0$ is brought up to process temperature (so that the reactions start) at the same time as the in- and outflows are started. Rather than considering the steady-state RTD of inflowing particles like Danckwerts did, Fan et al. set the residence time of all particles present in the vessel at $t = 0$ ("inventory particles") to zero, and consider their RTD, $F_{\Theta_1}(t)$, on this basis, in addition to considering the RTD, $F_{\Theta_2}(t)$, of the particles flowing in from $t = 0$ ("inflow particles"). We shall see that their analysis of unsteady state RTD leads to an intuitively somewhat obvious result. The main value of the analysis is in formulating a stochastic framework for modeling RTD.

In principle the RTD functions for both inflow and inventory particles vary with time during start-up, since the flowpattern in the vessel will change with time. The assumption is, however, that a fluid dynamical steady-state sets in quickly, so that this variation can be neglected.

Stochastic balances over the system elucidate the relationship between the probability functions.

For instance a balance over the vessel says that the N particles in the vessel at time t consist of:

$$N = N\big(1 - F_{\Theta_1}(t)\big) + n \int_0^t 1 - F_{\Theta_2}(t - \eta)d\eta, \tag{4.25}$$

where the first term on the RHS represents the inventory particles still in the system at t. In the second term $n(1 - F_{\Theta_2}(t - \eta)d\eta)$ is the fraction of inflow particles entering the vessel between times η and $\eta + d\eta$ that still remain in the vessel at t, so that the whole of the second term is the number of particles entering between 0 and t, and remaining in the vessel at t.

A balance between in- and outflows gives:

$$n\Delta t = N\big(F_{\Theta_1}(t + \Delta t) - F_{\Theta_1}(t)\big) + n \int_0^{t+\Delta t} F_{\Theta_2}(t + \Delta t - \eta)d\eta - n \int_0^t F_{\Theta_2}(t - \eta)d\eta, \tag{4.26}$$

where n is the number of particles entering per unit time and N the number in the vessel; all particles have the same volume. The LHS represents the inflow during $(t, t + \Delta t]$, first term on the RHS the outflow of inventory particles during $(t, t + \Delta t]$, second term the particles that flowed in during the period $[0, t + \Delta t]$ and have left at $t + \Delta t$ and the last term the particles that flowed in during the period $[0, t]$ and have left at t. The difference of the last two terms is thus the inflow particles leaving during $(t, t + \Delta t]$. To see this more clearly, please refer to the timeline in Figure 4.5

Figure 4.5: Timeline

Dividing both sides by $n\Delta t$ and taking the limit as $\Delta t \to 0$, and using the definition of the differential quotient leads to:

$$\frac{N}{n} f_{\Theta_1}(t) + \int_0^t f_{\Theta_2}(t-\eta)d\eta = 1. \tag{4.27}$$

Fan et al. also derive expressions for the *transient internal and exit age distributions*. We discuss the exit age distribution here.

Denoting, as before, the exit age distribution with the letter E, $E_t(\zeta)\Delta\zeta$ denotes the expected fraction of exiting particles with residence time, or age, between ζ and $\zeta+\Delta\zeta$ at time t. The expected number of such particles is:

$$\begin{aligned} &n\Delta t E_t(\zeta)\Delta\zeta \\ &= \sum_{i=1}^{N\left(F_{\Theta_1}(t+\Delta t)-F_{\Theta_1}(t)\right)} U(\zeta_{1i}-\zeta) - U\big(\zeta_{1i}-(\zeta+\Delta\zeta)\big) \\ &+\sum_1^j \left(\sum_{i=1}^{n\Delta\eta\left(F_{\Theta_2}(t+\Delta t-\eta)-F_{\Theta_2}(t-\eta)\right)} U(\zeta_{2i}-\zeta) - U\big(\zeta_{2i}-(\zeta+\Delta\zeta)\big)\right). \end{aligned} \tag{4.28}$$

Here the first term on the RHS counts the inventory particles leaving during $(t, t+\Delta t]$ (the upper limit of the sum) and having residence times, ζ_{1i}, between ζ and $\zeta+\Delta\zeta$. The summand is in terms of the unit step function:

$$U(t-\theta) = \begin{cases} 0 \text{ for } t < \theta \\ 1 \text{ for } t \geq \theta \end{cases}$$

and thus equals 1 if the inventory particle exiting has an age between ζ and $\zeta+\Delta\zeta$ and 0 otherwise. In the second term on the RHS, the inner sum counts the inflow particles entering during $(\eta, \eta+\Delta\eta]$ and leaving during $(t, t+\Delta t]$ (given by the upper limit of the sum), and having age between ζ and $\zeta+\Delta\zeta$ (counted by the summand like before). The outer sum is summing over all intervals $(\eta, \eta + \Delta\eta]$. The entire term thus counts all inflow particles exiting between t and $t + \Delta t$ and having ages between ζ and $\zeta + \Delta\zeta$.

Again dividing through by Δt and $\Delta\zeta$ and taking the limit as both go to zero gives:

$$\begin{aligned} E_t(\zeta) &= \frac{N}{n} f_{\Theta_1}(t)\delta(t-\zeta) + f_{\Theta_2}(\zeta)U(t-\zeta) \\ &= \big(1 - F_{\Theta_2}(t)\big)\delta(t-\zeta) + f_{\Theta_2}(\zeta)U(t-\zeta), \end{aligned} \tag{4.29}$$

where we have used the definitions of differentials and of Dirac's delta function. The last result can be seen as follows: $nF_{\Theta_2}(t)$ is the total flowrate out of inflow particles, so that $n\big(1 - F_{\Theta_2}(t)\big)$ is the flowrate out of inventory particles, which is also equal to $N\big(dF_{\Theta_1}(t)/dt\big) = Nf_{\Theta_1}(t)$.

Equation (4.29) is a statement of the rather intuitive result that the outflow at time t consists of a (diminishing) fraction of inventory particles with residence time t, and an (increasing) fraction of inflow particles with Danckwerts' steady-state RTD.

To obtain an expression for the steady-state case, we allow t to go to infinity, then $\delta(t-\zeta) \to 0$ and $U(t-\zeta) \to 1$, so that:

$$E_{t\to\infty}(\zeta) = E(\zeta) = f_{\Theta_2}(\zeta). \tag{4.30}$$

Fan et al. formulate a *Markov process for the particle lifetime* as a "pure-death" process both for transient and steady-state situations. In this, the entrance of a particle is a deterministic event, and the exit is governed by a probability rate, $\mu_2(t)$. For the steady state case:

$$\frac{dF_{\Theta_2}(t)}{dt} = \big(1 - F_{\Theta_2}(t)\big)\mu_2(t). \tag{4.31}$$

Solving this gives:

$$F_{\Theta_2}(t) = 1 - \exp\left(-\int_0^t \mu_2(s)ds\right), \tag{4.32}$$

which, together with Equation 4.30 gives:

$$E(t) = \mu_2(t)\exp\left(-\int_0^t \mu_2(s)ds\right). \tag{4.33}$$

They discuss the two ideal cases of pure plug flow, PF, and a single CSTR, in the former:

$$\mu_2(t) = \begin{cases} 0 & \text{for} \quad t \le \frac{N}{n} \\ \infty & \text{for} \quad t > \frac{N}{n} \end{cases}, \tag{4.34}$$

and in the latter:

$$\mu_2(t) = \mu_2 = \frac{n}{N}. \tag{4.35}$$

Thus in a CSTR any particle has a constant probability rate of exiting the vessel, independent of its history. We note that N/n is the same as V/Q in the previous sections.

4.3 Danckwerts' Law for Mean Residence Time Revisited

We devote the following section to elucidating Danckwerts' law for the mean residence time using a stochastic approach. This work was first presented in Gottschalk et al.[52].

4.3.1 Introduction

Satisfying Danckwerts' law for mean residence time is often a challenge, for instance for numerical solutions of the convection-diffusion equation. The boundary conditions used may induce errors either in the mean residence time or in the concentrations in the interior or at the boundaries of the domain (see, for instance, [51, 115]).

The object of these sections is to examine the conditions required for Danckwerts' law for mean residence time to be satisfied. To do this, a stochastic Markov model for the position of a particle in a system with through-flow is formulated, and the mathematical requirements for satisfaction of Danckwerts' law examined.

4.3.2 Mathematical Model

We consider the position of a particle or fluid element in a vessel of volume V' with volumetric inflow and outflow Q. Both space and time are discretized, each spatial cell has volume $\Delta V'$, and the length of a time interval is $\Delta t'$. We use these two magnitudes as scales for volume and time, so that the dimensionless volume of the vessel is $N = V'/\Delta V'$, the dimensionless time is $t = t'/\Delta t'$, and the dimensionless residence time $T = T'/\Delta t'$. The dimensionless volumetric throughflow becomes $v = Q\Delta t'/(\Delta V')$. Necessarily $v \leq 1$, which imposes the restriction $\Delta t' \leq \frac{\Delta V'}{Q}$ on the length of the time interval.

The particle's position is described by a random variable, $(X_n)_{n\geq 0}$, and space is discretized so that the state space for X_n is finite: $\{1, 2, 3, \ldots, N+1\}$. We let the process of particle movement be Markovian, and denote the transfer probabilities $p_{ij} := P(X_n = j|X_{n-1} = i)$. The p_{ij} constitute an $(N+1) \times (N+1)$ transfer matrix $\mathbf{P}$.

The probability distribution of the particle's position models the concentration profile of a pulse of inifinitely many non-interacting particles introduced in an inlet cell with index 1 at time $t = 0$. The outlet is modeled by an absorbing cell with index $N+1$. We denote the $N \times N$ transfer matrix without the absorbing cell by $\mathbf{Q}$.

We note that the n-times transition probabilities, $p_{ij}^n := P(X_n = j|X_0 = i) = (\mathbf{P}^n)_{ij} = (\mathbf{Q}^n)_{ij}$ for $1 \leq i, j \leq N$, see also Equation (1.26). The last equality holds because the $(N+1)$'th cell is absorbing.

Now:

$$(\mathbf{I} - \mathbf{Q}) \sum_{n=0}^{k} \mathbf{Q}^n = \mathbf{I} - \mathbf{Q}^{k+1}, \tag{4.36}$$

where k is the number of time steps. That the absorbing state is reachable from every state implies that $\mathbf{Q}^{k+1} \to 0$ for $k \to \infty$, so that:

$$(\mathbf{I} - \mathbf{Q}) \sum_{n=0}^{\infty} \mathbf{Q}^n = \mathbf{I}. \tag{4.37}$$

Thus $\sum_{n=0}^{\infty} \mathbf{Q}^n = (\mathbf{I} - \mathbf{Q})^{-1}$.

The expected total residence time for the particle in cell j if it starts in cell i is:

$$E\left[\sum_{n=0}^{\infty} \mathbb{I}_{X_n=j} \middle| X_0 = i\right] = \sum_{n=0}^{\infty} (\mathbf{Q}^n)_{ij} = \left(\sum_{n=0}^{\infty} \mathbf{Q}^n\right)_{ij} = (\mathbf{I} - \mathbf{Q})_{ij}^{-1}, \tag{4.38}$$

where $\mathbb{I}_{X_n=j}$ is the indicator function for $X_n = j$. $\sum_{n=0}^{\infty} (\mathbf{Q}^n)_{ij}$ is the total probability for transference from i to j.

The time that the Markov chain is in the non-absorbing states, or the time-to-absorption, is: $T = \sum_{j=1}^{N} \sum_{n=0}^{\infty} \mathbb{I}_{X_n=j}$, the expected value of which is:

$$E[T|X_0 = i] = \sum_{j=1}^{N} E\left[\sum_{n=0}^{\infty} \mathbb{I}_{X_n=j} \middle| X_0 = i\right] = \sum_{j=1}^{N} (\mathbf{I} - \mathbf{Q})_{ij}^{-1}. \tag{4.39}$$

Now we impose the following conditions on the transfer probabilities:

1. $\sum_{j=1}^{N} p_{j,1} = 1 - v$

2. $\sum_{j=1}^{N} p_{j,i} = 1$ for all $2 \leq i \leq N$

3. $\sum_{j=1}^{N} p_{j,N+1} = v$.

Since the matrix $\mathbf{I} - \mathbf{Q}$ is invertible, the matrix equation:

$$(\mathbf{I} - \mathbf{Q})^t x = e_1, \tag{4.40}$$

where $e_1 := (1, 0, \ldots, 0)^t$ is an N-dimensional vector, has a unique solution, which, by the standard rules for solving matrix equations, is the first row of the matrix $(\mathbf{I} - \mathbf{Q})^{-1}$. By conditions 1 and 2 $\sum_{i=1}^{N} (\mathbf{I} - \mathbf{Q})^t_{1i} = v$ and $\sum_{i=1}^{N} (\mathbf{I} - \mathbf{Q})^t_{ji} = 0$ $(2 \leq j \leq N)$, so that the N-dimensional vector: $x_0 := \frac{1}{v}(1, \ldots, 1)$ is a solution to the Equation (4.40).

By Equation (4.39), the expected residence time in the non-absorbing states, which constitute the vessel is:

$$E[T|X_0 = 1] = \sum_{j=1}^{N} (\mathbf{I} - \mathbf{Q})^{-1}_{1j} = \frac{N}{v}. \tag{4.41}$$

4.3.3 Physical Ramifications

The N non-absorbing cells constitute the vessel, and the absorbing cell $N + 1$ the outlet. Equation (4.41) is therefore a statement of Danckwerts' law for the mean residence time.

The probability of the particle occupying a given cell, $p(n, i)$ can be seen as the fraction of a pulse of marked material in that cell. The transition probabilities times the position probability, $p_{ij}p(n, i)$ can be seen as the flow of marked material from cell i to cell j during time step n.

The requirement that the absorbing state (or cell) must be reachable from all other states translates to the requirement that there must be no zones in the vessel that are completely "dead".

The most interesting about the above analysis is the conditions that are not made.

None of the above depends on the *form* of $\mathbf{P}$, and the analysis can therefore be applied to any stacking of cells in 1, 2 or 3 dimensions, and any distribution of material movement in the vessel. Specifically as long as the outlet is reachable from all cells, the degree of "deadness" of partially dead regions, such as those discussed by Danckwerts[30] in connection with "hold-back" has no influence on the mean residence time.

Conditions 1, 2 and 3 are simply volume conservation within the vessel. Condition 2 says that the volume that flows into a cell (marked and unmarked material) is the same as what flows out, and conditions 1 and 3 that the volume flowing into the reactor is the same as that which flows out.

The above also implies Gibilaro's result expressed in Equation (4.16). By Equation (4.38) the expected residence time in cell j, when starting in cell 1, is $(\mathbf{I}-\mathbf{Q})^{-1}_{1j}$. Since we found that the first row of $(\mathbf{I}-\mathbf{Q})^{-1}$ is given by $x_0 := \frac{1}{Q}(1,\ldots,1)$, the expected residence time in each cell is $\frac{1}{Q}$. Physically we can make cells arbitrarily small, and combine an arbitrary number, N_s, of them to create any subregion of volume $\frac{N_s}{N}$, for which the expected residence time is thus $\frac{N_s}{Q}$, which is Gibilaro's result.

4.4 Stochastic Modeling of RTD for Complex Systems

This section is dedicated to stochastic modeling of the RTD in systems consisting of different vessels with a complex pattern of flows between them, such as one may find in a processing plant.

To our knowledge Gibilaro et al. [49] pioneerd a stochastic approach based on Markov chains for modeling the residence time distribution in complex systems in process technology. Subsequently, in the middle of the 1980's, a series of papers were published by two groups wherein residence time distribution in complex flow systems were modeled using stochastic techniques. One of the groups was Rubinovitch and Mann [109–111], and the other Fan and co-workers [44].

We start by giving an account of the pioneering work of Gibilaro et al. [49]. Although this is a model for a single processing vessel, this vessel is modeled as if it consisted of a number of interconnected vessels, and we need to start with this work in order to account for the historical development of the methodology. As an example of the papers in the 80's, we take the excellent paper by L. T. Fan et al. [44], wherein the residence time distribution of molecules or particles (sometimes only referred to as "particles") in complex flow systems are derived using a stochastic approach based on Markov chains. Finally we mention the main differences between the treatments of Fan et al. and Rubinovitch and Mann, without giving a detailed account of the latter.

4.4.1 Pioneering Work of Gibilaro et al.

Gibilaro et al. [49] propose to solve a model for the mixing and residence time distribution in a stirred tank reactor by a stochastic model based on Markov chains. The model for the flow pattern in the reactor, due to van de Vusse [123], is relatively simple, yet too complex for analytical solution methods. The model of van de Vusse may have been superceded (see, for instance, Section 6.1.1), but we are here interested in the innovative use of stochastic modeling in process technology introduced by Gibilaro et al.

To understand the model concept of Gibilaro et al., we briefly describe the physical phenomena taking place in continuous stirred tank reactors, or CSTRs. We shall return to the issue of the flow pattern in (semi-) batch stirred tank reactors in Section 6.1.1. A physical CSTR contains one or more impellers, often a single impeller mounted axially as indicated in Figure 4.6. The rotation of the impeller creates a recirculatory flow pattern in the tank as indicated. The impeller may have either flat blades, as in Figure 4.6A, which will be outward pumping and give rise to the flow pattern indicated, or be have inclined blades, as in Figure 4.6B, which may be upward or downward pumping. The recirculatory flowpattern generated by a downward pumping impeller is indicated in Figure 4.6B.

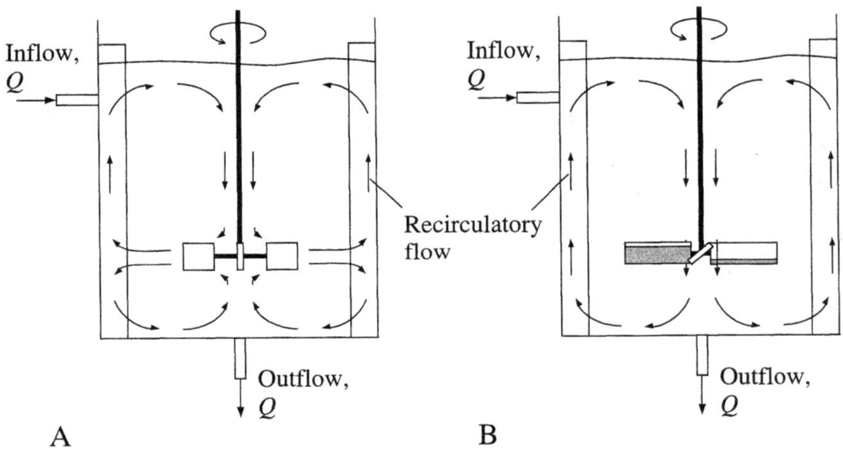

Figure 4.6: Two possible configurations of a continuous stirred tank reactor, or CSTR. **A** Impeller with flat blades pumping outward. **B** Impeller with inclined blades pumping downward. The in- and outlet flows may each be distributed almost arbitrarily over one or more positions

Gibilaro et al. modeled a stirred tank reactor as a number of compartments of equal volume arranged around the stirrer, and exchanging material with it. Figure 4.7 shows the two arrangements considered. In these the central point represents the stirrer, Q is the through-flow, and r is the recirculation flow in each of three flowloops.

The model is a discrete time Markov process, where the position of

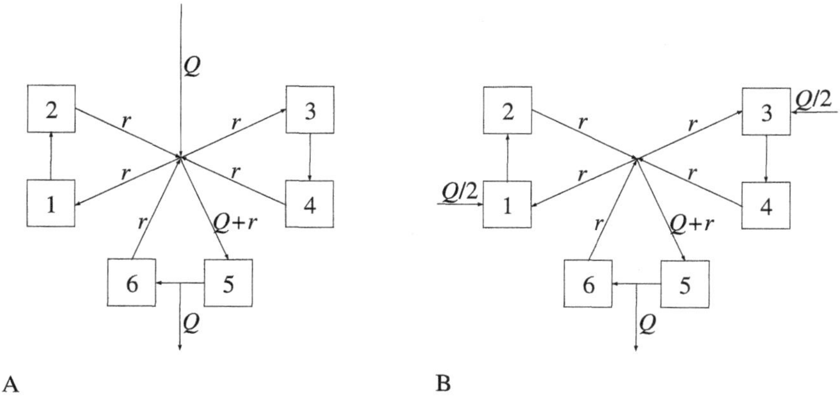

Figure 4.7: Sketches showing the two systems considered by Gibilaro et al. [49] to model the flowpattern in a continuous stirred tank reactor. If q is the total recirculatory flow due to the pumping action of the impeller, $r = q/3$

a molecule or particle is described by a stochastic variable $X_i(n)$, say, with transition probability matrix $\mathbf{P}$ having elements $p_{i,j}$ and state space $i = 1, \ldots, N+1$, where the $(N+1)$'st compartment is an absorbing one[2], and represents the exterior of the reactor, such that $p_{N+1,N+1} = 1$. They define a probability vector, which we will here call $\mathbf{p}(n)$, with elements $p_i(n)$.

Gibilaro et al. assumed all the compartments to have the same volume and, as can be seen from Figure 4.7 that the flow through them all is the same. They state, however, as we shall also see in Section 4.4.2 below, that any other assumptions about the volumes and the flows could be handled equally easily.

If each of the vessels are stirred vessels (CSTRs) the probability of a particle remaining in the compartment, p_{ii}, depends on the length chosen for the time interval, Δt, and, in accordance with Equation (A.3), is:

$$p_{ii} = e^{-\Delta t/\tau_i}, \tag{4.42}$$

[2]The absorbing compartment is given the index N in the paper of Gibilaro et al., but is called $N+1$ here for consistency

while the probability of transferring to another compartment is:

$$\sum_{j \neq i} p_{ij} = 1 - e^{-\Delta t/\tau_i}, \tag{4.43}$$

where τ_i is the mean residence time in the compartment, equal to its volume divided by the volumetric throughflow, r/V_i, in accordance with Danckwerts' law, Equation (4.8), where we call the compartment volume V_i here.

For system A in Figure 4.7, the full transfer probability matrix between specific compartments via the impeller can, by simple flow balances between the compartments, then be determined to be:

$$\begin{aligned}
p_{21} &= p_{23} = p_{41} = p_{43} = p_{61} = p_{63} = \frac{r}{3r+Q}(1 - e^{(-r/V_i)\Delta t}),\\
p_{25} &= p_{45} = p_{65} = \frac{r+Q}{3r+Q}(1 - e^{(-r/V_i)\Delta t}),\\
p_{56} &= \frac{r}{r+Q}(1 - e^{(-(r+Q)/V_i)\Delta t}),\\
p_{57} &= \frac{Q}{r+Q}(1 - e^{(-(r+Q)/V_i)\Delta t}),\\
p_{11} &= p_{22} = p_{33} = p_{44} = p_{66} = e^{(-r/V_i)\Delta t)}\\
p_{55} &= e^{(-(r+Q)/V_i)\Delta t)},\\
p_{77} &= 1.
\end{aligned} \tag{4.44}$$

All other transfer probabilities are zero.

The transfer probabilities may be rewritten in terms of dimensionless quantities by defining a dimensionless time step: $\Delta T = \Delta t(Q/V)$, where V is the total volume of the six transient compartments, and substituting V_i with $V/6$ and r with $q/3$. For instance:

$$\frac{r}{3r+Q}(1 - e^{(-r/V_i)\Delta t}) \text{ becomes: } \frac{q/Q}{3(q/Q+1)}(1 - e^{(-2q/Q)\Delta T}).$$

All the other transfer probabilities may be similarly written in terms of the ratio of internal circulation flow to throughflow, q/Q, showing that it is sufficient to know this ratio to quantify the entire transfer probability matrix.

Gibilaro et al. compare their model with experimental data, and find that the match is good. We will not show any results of this model here, but proceed directly to a more general model for the RTD in complex flow systems.

4.4.2 Stochastic RTD model of Fan et al.

Fan et al. consider a system of compartments, which play a role equivalent to that of the spatial cells in the previous Markov chain models in this book. An example of such a process is given in Figure 4.8.

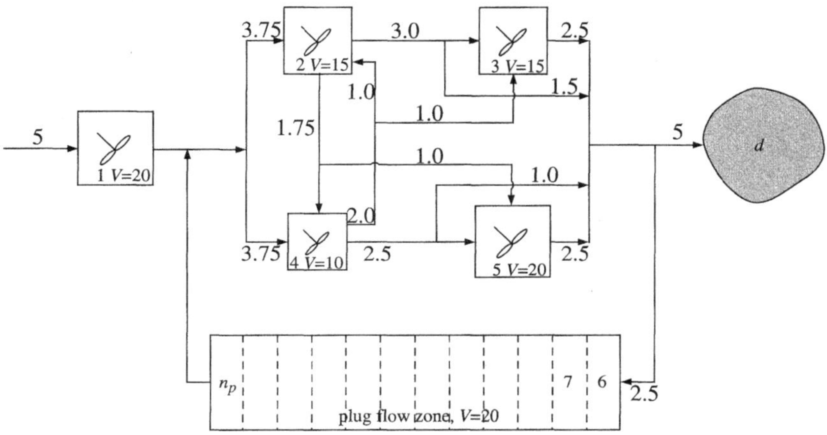

Figure 4.8: Example system considered by Fan et al.

They model the particle's position on the system, i.e. in which compartment it is situated, as a stochastic variable X_n with state space $\{1, 2, 3, \ldots, N+1\}$ and allow the stochastic variable to undergo a Markov chain, which is thus discrete in space and is also discrete in time. The transfer probability matrix, $\mathbf{P} = (p_{ij})_{1 \le i,j \le N+1}$ is quantified in accordance with the exchange rates between the compartments. These exchange rates are considered as material flows, and they are balanced in advance, so that the conditions for validity of Danckwerts' law derived in section 4.3 are satisfied *a priori*.

Three types of compartments are considered (see Figure 4.8):

- Compartments acting as continuous stirred vessels (CSTRs). These are treated in a similar way to the compartments in the model of Gibilaro et al [49]. The probability of a particle remaining in the compartment, p_{ii} is:

$$p_{ii} = e^{-\Delta t/\tau_i}, \tag{4.45}$$

where τ_i is the mean residence time in the compartment, equal to its volume divided by the volumetric throughflow V_i/Q_i. The probability of flowing to another compartment, j, is then, as by Gibilaro et al., found from:

$$p_{ij} = \frac{Q_{ij}}{\sum_{j \neq i}^{N+1} Q_{ij}} (1 - p_{ii}), \tag{4.46}$$

where Q_{ij} is the flowrate from compartment i to compartment j.

- Compartments in a plug flow zone. Plug flow zones are split in exactly so many compartments, n_p, that the material residence time in a compartment, τ_c, equals Δt. Thus:

$$\Delta t = \tau_c = \frac{V_c}{Q_p} = \frac{V_p/n_p}{Q_p} = \frac{\tau_p}{n_p} \Rightarrow n_p = \frac{\tau_p}{\Delta t}, \tag{4.47}$$

where V_c and V_p are the compartment and plug flow zone volumes, respectively, and τ_p the residence time in the plug flow zone[3]. This gives transfer probabilities in such compartments of:

$$p_{ij} = \begin{cases} 1 \text{ if } j = i + 1, \\ 0 \text{ otherwise.} \end{cases} \tag{4.48}$$

- An absorbing compartment, which is given the index $N+1$ elsewhere in this book, but is given the index d in this section. The transfer probabilities are:

$$p_{dj} = \begin{cases} 1 \text{ if } j = d, \\ 0 \text{ otherwise.} \end{cases} \tag{4.49}$$

Let θ_{ij} denote the number of time steps it takes for a particle starting in compartment i to first reach compartment j, and $f_{ij}(n)$ denote $P(\theta_{ij} = n) = P(X_n = j, X_m \neq j (m = 1, 2, \ldots, n-1) | X_0 = i)$. When $i = j$, $f_{ii}(n)$ is the probability that the first recurrence in cell i is at time n.

As before, let $p_{ij}(n)$ be the multi-step transition probability i to j: $P(X_n = j | X_0 = i)$. If the particle is in j at time n, it may have been there earlier, for instance at time m. The p_{ij} therefore relates to the f_{ij} as:

$$p_{ij}(n) = \sum_{m=1}^{n} f_{ij}(m) p_{jj}^{(n-m)} = \sum_{m=1}^{n-1} f_{ij}(m) p_{jj}^{(n-m)} + f_{ij}(n), \ n = 1, 2, \ldots$$

[3]We note that this choice of n_p is a prerequisite for obtaining true plug flow. In Chapter 3 we used a non-zero transfer probability against the flow direction to simulate dispersion. But even in the absence of such a transfer probability against the flow, any other choice of n_p than that in Equation (4.47) will lead to some dispersion of the probability distribution. For example for $n_p < \tau_p/\Delta t$ both p_{ii} and $p_{i,i+1}$ would have to be non-zero to obtain the correct flow through the plug flow zone

(4.50)

This constitutes a recursive relation for computing successive values of $f_{ij}(n)$:

$$f_{ij}(n) = \begin{cases} 0, & n = 0 \\ p_{ij}, & n = 1 \\ p_{ij}^{(n)} - \sum_{m=1}^{n-1} f_{ij}(m) p_{jj}^{(n-m)}, & n = 2, 3, \ldots \end{cases} \tag{4.51}$$

If we put $j = d$, f_{id} is the probability of a particle starting in i having reached the absorbing cell precisely at time n. This probability can also be seen as the fraction of exiting material that resided in the system between $n\Delta t$ and $(n + 1)\Delta t$ seconds, which in turn, as seen by comparing with Equation (4.4), approximately[4] equals $E(n\Delta t)\Delta t$, where $E(t)$ is the exit age distribution of Danckwerts. Thus:

$$E_i(n\Delta t) \approx \frac{f_{id}(n)}{\Delta t}. \tag{4.52}$$

where E_i is the E-function given that the particle started in compartment i. Note also that p_{dd} is unity.

Another way of relating this process to Danckwerts' functions is to realize that $F(t)$ is the probability of the particle having reached the absorbing compartment at time t, and this is given by the component $p(n, d)$ of the probability vector $\mathbf{p}(n)$ describing the particle's position at the n'th time step:

$$\mathbf{p}(n) = \mathbf{p}(0)\mathbf{P}^n, \tag{4.53}$$

as we will also point out in Equation (5.2) of Chapter 5. Please compare with Equation (3.16), as well. In this case, where the particle starts in compartment i, the definition of the multi-step transfer probabilities as: $p_{ij}^{(n)} := P(X_n = j | X_0 = i)$ shows that $p(n, d) = p_{id}^{(n)} \approx F_i(n\Delta t)$.

Figure 4.9 shows both the E_i and the F_i functions for the highly complex system in Figure 4.8 plotted against the dimensionless time, $\theta = t/\tau$, where τ is the mean residence time in the system as a whole, namely the total volume of all the compartments divided by the flow in and out of the system.

This is a clear illustration of the power of the stochastic modeling technique, compared to conventional modeling based on conservation equations

[4]approximately because the process has been discretized

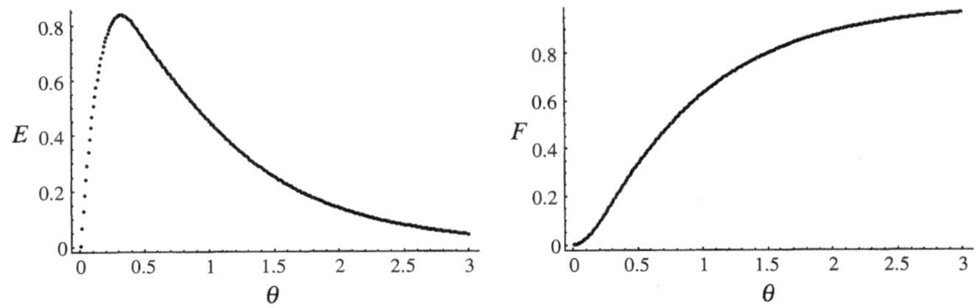

Figure 4.9: E- and F-curves for the complex system shown in Figure 4.8

and explicit mathematical expressions. The latter method would not yield any solution for this highly complex system, which contains CSTRs of various volumes and a plug flow zone, all connected in a complicated way. Also Monte-Carlo type simulations to obtain the E- and F-curves would be complex and time-consuming, while the present method is evaluated in seconds with matrix manipulations. In spite of its complexity, a system like the one illustrated in the figure is by no means atypical for what one may find in the processing industry.

The above analysis was based on the particle being in one compartment, i, initially, corresponding to a single inlet to the process. This will cover many cases in practice. However, a system may have multiple inlets, and Fan et al. model this with a distributed initial probability vector for the particle, $\mathbf{p}(i, 0) = Q_i/Q$, where Q_i is the flowrate into the i'th compartment. The E-function then becomes:

$$E(n\Delta t) = \sum_{i=1}^{N} p(0, i) E_i(n\Delta t). \tag{4.54}$$

Let us consider one further simple, but illustrative, example of the power of this method for RTD modeling in the processing industry. A problem in industry is often the occurrence of dead-zones, or zones that are almost dead in a processing vessel. A significant part of the throughflow may "short-circuit" and take place through a localized, highly permeable region, while the flow through other zones ("deadzones") may be very slow[5]. We can study this phenomenon with the relatively simple system illustrated in Figure 4.10.

[5]or non-existent if the zone is completely dead; this is, however a trivial case, since the zone then no longer enters the analysis at all

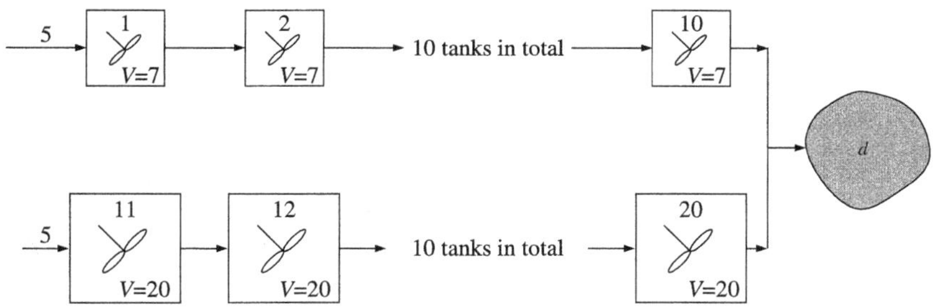

Figure 4.10: A simple flow system to study the effect of near-dead zones

Each row of CSTRs represents a path for the fluid through the process. There are 10 CSTRs in each path, representing moderate mixing as discussed earlier in this chapter. The top row are small vessels, and represent the short-circuit path while the bottom row represent the "deadzones". The distribution of the flow between the rows, and the number and volume of the CSTRs may, of course, be adjusted to suit the process under consideration.

The system is much simpler than the one in Figure 4.8, and very simple when considering the highly complex systems amenable to analysis by this method.

With the compartment numbering and the flows and volumes indicated in the figure, Equations (4.45) and (4.46) give the following transfer probability matrix:

For $i \leq 9$:

$$p_{ij} = \begin{cases} 0.8669 \text{ if } j = i, \\ 0.1331 \text{ if } j = i + 1, \\ 0 \text{ otherwise.} \end{cases}$$

For $i = 10$: $p_{ii} = 0.8669$, $p_{id} = 0.1331$ and $p_{ij} = 0$ otherwise.

For $11 \leq i \leq 19$:

$$p_{ij} = \begin{cases} 0.9512 \text{ if } j = i, \\ 0.0488 \text{ if } j = i + 1, \\ 0 \text{ otherwise.} \end{cases}$$

For $i = 20$: $p_{ii} = 0.9512$, $p_{id} = 0.0488$ and $p_{ij} = 0$ otherwise.

In the simple system considered here there is no exchange of material between the two rows of compartments, although in principle there is no problem in implementing material exchange between them. We consider, as indicated in the figure, an inlet stream evenly distributed between compartments 1 and 11, which in the model translates to a distributed initial probability vector of $\mathbf{p}(0) = (0.5, 0, \ldots, 0, 0.5, 0, \ldots, 0)$. We solve the two flowconfigurations corresponding to the two inlets separately, and combine the solutions afterwards using Equation (4.54). Note that only one of the rows come into play in each of these two solutions, since they are not exchanging material.

Figure 4.11 shows the E and the F-curves for this system. The characteristic bimodal E-curve for a system with deadzones is clearly recognizable. The F-curve does not show the effect so clearly, as is the wont with F-curves, but a point of inflexion is visible.

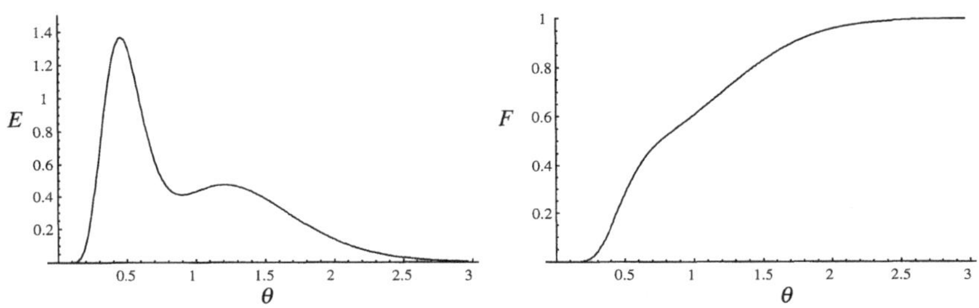

Figure 4.11: E and F-curves for the system in Figure 4.10

Fan et al. continue in their paper to derive expressions for the mean and variance of the particle residence time. We call the random variable denoting the total residence time of a particle starting in compartment i Z_i. To give an account of the derivation of the mean and variance of Z_i, we can make use of some of the material in Section 4.3, since the class of systems considered there also includes the one considered by Fan et al.

In fact, we have already found an expression for the mean of Z_i. In Equation (4.38) we found that the expected total dimensionless residence time in compartment j of a particle starting in i was the ij component of the matrix $(\mathbf{I} - \mathbf{Q})^{-1}$, and thus the expected value of Z_i, is:

$$E[Z_i] = \sum_{j=1}^{N} (\mathbf{I} - \mathbf{Q})_{ij}^{-1}, \text{ or in matrix form } [E[Z_i]] = (\mathbf{I} - \mathbf{Q})^{-1}\boldsymbol{\xi} \quad (4.55)$$

(compare with Equation (4.39)), where we have denoted the column vector of $E[Z_i]$ by $[E[Z_i]]$, $\mathbf{Q}$ is the transition probability matrix between the non-absorbing states, $\mathbf{I}$ is the identity matrix and $\boldsymbol{\xi}$ is a columnvector with components of unity.

We now wish to determine the variance of Z_i from the well-known formula: $Var[Z_i] = E[Z_i^2] - E[Z_i]^2$, and for this we need $E[Z_i^2]$. To find this, Fan et al. consider that a particle starting in i may either, with probability p_{id}, proceed directly to d, in which case $Z_i = 1$, or, with probability p_{ij}, to some other (non-absorbing) compartment j in which case $Z_i = Z_j + 1$. Thus we can rewrite $E[Z_i^2]$:

$$\begin{aligned} E[Z_i^2] &= p_{id} + \sum_{j=1}^{N} p_{ij} E[(Z_j + 1)^2] \\ &= p_{id} + \sum_{j=1}^{N} p_{ij} (E[Z_j^2] + 2E[Z_j] + 1) \\ &= \sum_{j=1}^{N} p_{ij} (E[Z_j^2] + 2E[Z_j]) + 1 \end{aligned} \tag{4.56}$$

where we have also made use of the fact that $p_{id} + \sum_{j=1}^{N} p_{ij} = 1$.

We can write Equation (4.56) in matrix form:

$$[E[Z_i^2]] = \mathbf{Q}[E[Z_i^2]] + 2\mathbf{Q}[E[Z_i]] + \boldsymbol{\xi}. \tag{4.57}$$

Rearranging:

$$[E[Z_i^2]] = (\mathbf{I} - \mathbf{Q})^{-1}(2\mathbf{Q}[E[Z_i]] + \boldsymbol{\xi}).$$

Substituting for $[E[Z_i]]$ using (4.55) and rearranging gives:

$$[E[Z_i^2]] = \left(2(\mathbf{I} - \mathbf{Q})^{-1} - \mathbf{I}\right)(\mathbf{I} - \mathbf{Q})^{-1}\boldsymbol{\xi}. \tag{4.58}$$

The columnvector of the variances the Z_i, $[Var[Z_i]]$, then becomes:

$$[Var[Z_i]] = \left(2(\mathbf{I} - \mathbf{Q})^{-1} - \mathbf{I}\right)(\mathbf{I} - \mathbf{Q})^{-1}\boldsymbol{\xi} - [(E[Z_i])^2]. \tag{4.59}$$

where the components of the columnvector $[(E[Z_i])^2]$ can be found from Equation (4.55).

In this way Fan et al. have determined the mean and variance of the dimensionless residence time for a particle starting in cell i in terms of the transfer probability matrix for the Markov process. The dimensional residence time has expected value, $E[T_i]$, of $E[Z_i]\Delta t$ and variance, $Var[T_i]$ of $Var[Z_i](\Delta t)^2$.

4.4.3 Work of Rubinovitch and Mann

In a series of three papers, also Rubinovitch and Mann [109–111] study the dynamics of a single particle in a complex flow structure as a stochastic variable, undergoing a Markov process discrete in time with an associated transfer probability matrix $\mathbf{P}$ with elements p_{ij}.

The type of system studied by Rubinovitch and Mann is somewhat different from that of Fan et al. above. An example system studied in the first of their papers [109] is shown in Figure 4.12.

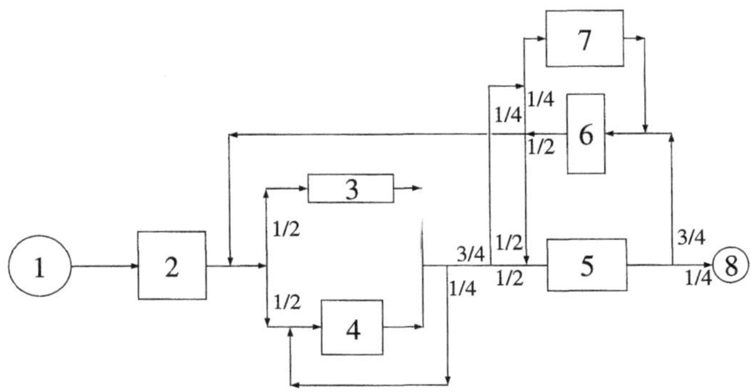

Figure 4.12: Example system considered by Rubinovitch and Mann [109]

Only transitions *between* the flow regions are considered as indicated. The inherent assumption is one transition per "time step", and they do not allow for a residence time in a given region by assigning a non-zero value for p_{ii}, such as was done in Equations (4.42) and (4.45). The transition matrix for the system shown in Figure 4.12 is:

$$\begin{pmatrix} 0 & 1 & 0 & 0 & 0 & 0 & 0 & 0 \\ 0 & 0 & 1/2 & 1/2 & 0 & 0 & 0 & 0 \\ 0 & 0 & 0 & 1/4 & 3/8 & 0 & 3/8 & 0 \\ 0 & 0 & 0 & 1/4 & 3/8 & 0 & 3/8 & 0 \\ 0 & 0 & 0 & 0 & 0 & 3/4 & 0 & 1/4 \\ 0 & 0 & 1/8 & 1/8 & 1/2 & 0 & 1/4 & 0 \\ 0 & 0 & 0 & 0 & 0 & 1 & 0 & 0 \\ 0 & 0 & 0 & 0 & 0 & 0 & 0 & 1 \end{pmatrix}.$$

Not all the diagonal elements, p_{ii} are zero, there is a loop around region 4 giving a non-zero probability of transference back to the same region. Also

$p_{ii} = 1$ for the absorbing region, of course.

We will not give a full account of the contents of these three papers. The authors derive a number of useful relations for the expected number of passes of a particle or molecule through each region in the first paper [109].

In the second one [110] they relate the total residence time in a given region to the number of passes through that region, and account for how the compound residence time distribution for, say, n passes may be computed by nth-fold convolution of the single-pass residence time distribution function—which is assigned some form from knowledge of the character of the region, e.g. ideal mixing for a CSTR region—with itself. They also show how the flowrates through each of the flow regions may be calculated from the number of passes through that region.

In the final paper [111], the authors extend the method to considering more than one fluid in the system.

Many useful relations are derived in these papers. However, a process technologist is most likely to start out with knowing the flows and the vessel volumes of the process he or she is trying to model. With this in mind the method of Rubinovitch and Mann, restricted as it is to

- the particle or molecule exiting the region it occupied each step (albeit possibly transferring back into the region it came from), and
- calculation of flows between the regions from the computed number of passes through each region,

represents a less direct method for modeling a given process than that described in Section 4.4.2.

Chapter 5

RTD in Continuous Fluidized Beds

In the previous chapters, we looked at the particle transport in fluidized beds, and at the general issue of residence time for molecules or particles in processes. In this chapter we shall to an extent combine the two, and discuss models for the particle residence time in continuous fluidized beds, that is fluidized beds where particles are continually added to and withdrawn from the bed vessel. Such processes are very common in industry, for instance fluidized bed dryers, the catalyst regenerator bed in the FCC (fluidized catalytic cracking) process, and fluidized bed riser reactors.

5.1 Types of beds considered here

We have, in Section 3.1, already discussed in some detail the flow regimes in batch fluidized beds. Continuous fluidized beds may be operated in any of the fluidized regimes illustrated in Figure 3.1. Particle addition and withdrawal may in principle be at any point in the bed. Often addition and withdrawal are arranged to give some degree of uniformity over the bed cross-section, and the axial positions are removed from each other to avoid particles "short-circuiting" the bed.

In this chapter we will discuss models for two bed configurations, both of which are illustrated schematically in Figure 5.1:

- A bubbling fluidized bed with addition of particles to the top of the bed and withdrawal from the bottom.
- A fluidized bed operated in the transport regime (a "riser"), with addition of particles at the bottom and withdrawal from the top.

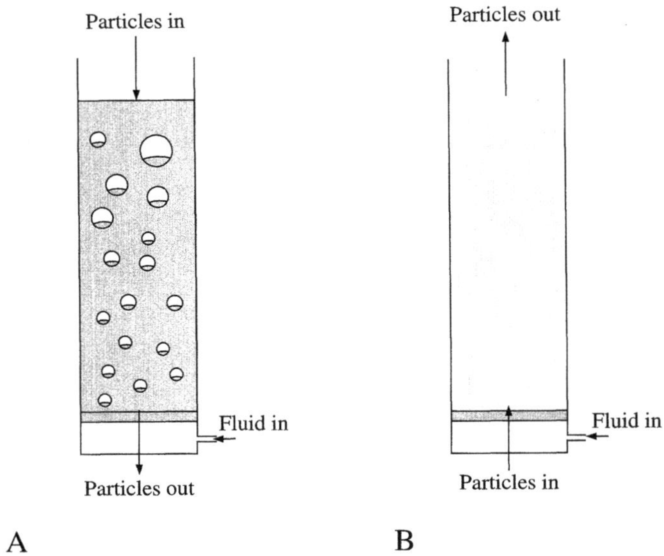

Figure 5.1: Continuous fluidized beds. A) A bubbling bed with particles added to the top and withdrawn at the bottom. B) A bed operated in the transport mode (a "riser"), with particles added to the bottom and withdrawn at the top

5.2 Bubbling bed

We formulate a stochastic model for the particle residence time in a continuous bubbling bed. This model was first published by Dehling et al. [39]. In contrast to the batch bubbling bed, we are here concerned with bulk particles, and are not considering the effects of segregation of dissimilar particles, although this latter topic is an interesting one, which may be addressed in the future.

5.2.1 Earlier work on particle RTD in bubbling fluidized beds

A number of studies in the scientific literature are dedicated to predicting the particle RTD in continuous fluidized beds. In most of this work attempts have been made to formulate semi-empirical models using the traditional tools of RTD theory. Some workers have used ideal mixers in series [77, 103], fitting the number of mixers to match predictions to experimental results. Another modeling approach has been based on plug flow with superimposed dispersion, where the dispersion coefficient was adjusted to fit experimental data [88, 122, 130].

However, a discrepancy between the model predictions and the actual behavior of continuous fluidized beds remained when using these simple approaches. Whittmann et al. [136] tried series of ideal mixers with reverse flow, and Heertjes, de Nie, and Verloop [59] tried a model incorporating combinations of ideal mixers in series and in parallel. Krishnaiah, Pydisetty, and Varma [80] proposed a model with a combination of a mixed section with a stagnant one and short circuiting.

A few studies have attempted to take the actual physical phenomena occurring in a continuous fluidized bed into consideration. Berruti et al. [7] proposed a model based on a series of compartments in the fluidized bed each with gross solid circulation. Morris et al. [94] presented a model based on the proposed existence of a velocity profile in the vertically moving fluidized solids, similar to the profile in a viscous fluid flowing in a pipe. Haines et al. [53] used the plug flow with axial dispersion approach, but they augmented their Fickian dispersion coefficient (which they assumed was caused mostly by the random collisional movement of individual particles) with a term accounting for extra dispersion caused by rising fluidization bubbles. Hoffmann and Paarhuis [64] showed by means of Monte-Carlo type simulations that the particle transport processes proposed for batch fluidized beds by Rowe and Partridge [108], and described in Chapter 3 in this book, could account also for the RTD of particles in continuous fluidized beds.

5.2.2 Particle transport mechanisms

We consider the same particle transport mechanisms as described in the bubbling bed sections of Chapter 3:

1. transport up in the wakes of fluidization bubbles,

2. a compensating transport down in the bulk and

3. dispersion due to the fluidization bubbles.

In fact, the main differences between the model considered in Chapter 3 and the one discussed below are that in the model below v_{seg} falls away, since we are not considering segregation of particles, but an additional downward velocity contribution in the bulk phase is present due to the through-flow of particles.

5.2.3 Model Formulation

The model is based on a Markovian model for the motion of individual particles through the bed. Denoting, as before, by X_t the vertical distance of the particle from the bed surface at time t, we model the process as a stochastic process (X_t).

We study a discrete model, obtained by dividing the reactor into N horizontal cells of equal width and modeling the particle's location at integer times only. The cells are numbered from top to bottom, with an extra cell with index $N + 1$ denoting the lower exit of the reactor (see Figure 5.2). Particles that have entered state $N + 1$ cannot return to the interior of the bed. The dynamics of the process is thus described by a Markov chain $(X_n)_{n \geq 0}$ with state space $\{1, 2, \ldots, N + 1\}$ and an absorbing boundary at $N + 1$. The process is basically a birth-death process, modified to allow for instant jumps to the first cell to model the possibility that a particle is caught in the wake of a rising fluidization bubble, very similar to the process described in Equations (3.12) and (3.13).

We show that this model gives rise to RTDs that capture the main features of empirically observed RTDs in fluidized beds. We show that the long tails of the RTD function are a consequence of the fact that the second-largest eigenvalue of the Markov transition matrix is very close to unity.

As in Chapter 3 empirical expressions from the literature for the particle flow in the wakes of rising gas bubbles are taken as basis for estimating the model parameters. For several parameter settings the model RTD is compared with experimental values. It turns out that the theoretically obtained RTD functions capture the main features of the experimental RTDs quite well.

In Dehling et al. [39] the study is extended to continuous models as limits of discrete Markov chains, obtained by letting space and time discretizations converge to zero in an appropriate way. The limit process is an ordinary diffusion process with reflecting/absorbing boundary conditions, modified to allow for instant jumps to the origin. A PDE for the particle density $p(t, x)$ in the continuous model is derived there, which then also provides the RTD function via $F(t) = 1 - \int_0^1 p(t, x)dx$. It is shown numerically that the particle density and the RTD function of the continuous model can be well approximated by the corresponding functions in the discrete model, provided the discretization is fine enough; for all practical purposes $N = 50$ cells turned out to be sufficient.

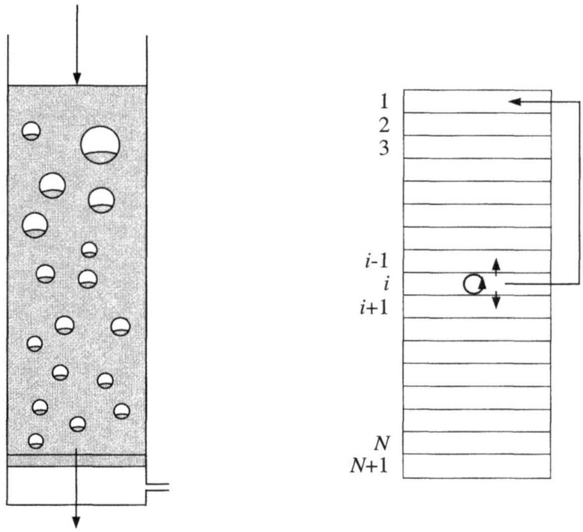

Figure 5.2: Diagram of a continuous bubbling fluidized bed and the discretized fluidized bed with inflow in the top cell and an absorbing cell $N+1$, symbolizing the exterior of the bed, at the bottom. The possible transitions in this model is that the particle can move one cell up, one cell down, remain in the cell, or move up to the top of the bed (caught in the wake of a fluidization bubble)

Discrete Markov model

In this section we set up the discrete model for particle transport in the fluidized bed. As for batch beds, we discretize space by subdividing the interior of the bed into N horizontal cells of equal width, labeled $i = 1, 2, \ldots, N$, but here we denote the lower exit of the bed with the index $i = N + 1$. Time is discretized as for batch beds. We denote the index of the cell that the particle visits at time n by X_n. We model the dynamics of the process by assuming that $(X_n)_{n \geq 0}$ forms a Markov chain, which is homogeneous in time. This Markov chain is fully specified once we know:

1. the probability vector $\mathbf{p}(0) = (p(0,1), \ldots, p(0, N+1))$ of the particle's initial position, where $p(0,i) = P(X_0 = i)$, and
2. the transition matrix $\mathbf{P} = (p_{ij})_{1 \leq i,j \leq N+1}$, where p_{ij} gives the conditional probability that the particle is in cell j at time $n+1$, given it

was in cell i at time n, i.e., $p_{ij} = P(X_{n+1} = j | X_n = i)$.

As for batch fluidized beds we define the probability vector $\mathbf{p}(n)$ with components $p(n,i) = P(X_n = i)$, and again we obtain the recursion formula for the probability distribution of the particle's position at time n:

$$p(n,j) = \sum_{i=1}^{N+1} p(n-1,i)p_{ij} \text{ in matrix notation: } \mathbf{p}(n) = \mathbf{p}(n-1)\mathbf{P}. \quad (5.1)$$

In what follows we always assume that the particle starts in the first cell; hence $p(0) = (1, 0, \ldots, 0)$. Then $p(n)$ is simply the first row of the nth power of the transition matrix.

From a macroscopic point of view, the probability vector $\mathbf{p}(n)$ gives the distribution of particles over the cells if at time $n = 0$ a large number of particles was placed in the first cell. In such a setup, (5.1) can be interpreted as a mass-balance equation.

Since a particle that has left the bed cannot return to the inside, the residence time equals the first exit time $T = \inf_{n \geq 0}\{X_n = N+1\}$. The residence time distribution is the distribution of T and can be described either by its cumulative distribution function $F(n) = P(T \leq n)$ or by its probability function $f(n) = P(T = n)$. As $T \leq n$ if and only if $X_n = N+1$, we have the simple identity

$$F(n) = P(T \leq n) = p(n, N+1), \quad (5.2)$$

and

$$f(n) = p(n, N+1) - p(n-1, N+1), \text{ for } n = 1, 2, \ldots$$

Our starting point is, as it was with batch fluidized beds, a conventional birth-death model, where a particle positioned at i can move one step ahead with probability β_i, one step back with probability δ_i, or stay in the present location with probability $\alpha_i = 1 - \beta_i - \delta_i$. We add to this the possibility of a complete return to the origin, whose probability is denoted by λ_i, thus modeling the chance that the particle gets caught in the wake of a gas bubble (see Figure 5.2). In this way we obtain the following transition probabilities for the interior of the bed, i.e., for $2 \leq i \leq N$:

$$\begin{aligned} p_{i,i} &= \alpha_i(1 - \lambda_i), \\ p_{i,i+1} &= \beta_i(1 - \lambda_i), \\ p_{i,i-1} &= \delta_i(1 - \lambda_i), \\ p_{i,1} &= \lambda_i, \end{aligned} \quad (5.3)$$

which are the same as those in Equation (3.12). Please note that for $i = 2$ the probability p_{21} is obtained by adding the first two probabilities, resulting in $\lambda_2 + \delta_2(1 - \lambda_2)$.

Regarding the boundaries, we assume reflection at the origin and absorption at $i = N + 1$, i.e.,

$$\begin{aligned} p_{1,1} &= 1 - \beta_1(1 - \lambda_1), \\ p_{1,2} &= \beta_1(1 - \lambda_1), \\ p_{N+1,N+1} &= 1. \end{aligned} \tag{5.4}$$

The last cell ($i = N + 1$) is identified with the exterior of the bed from where there is no return to the bed. The full transition matrix for this model is then:

$$\begin{pmatrix} 1-\beta_1(1-\lambda_1) & \beta_1(1-\lambda_1) & 0 & \cdots & 0 & 0 \\ \lambda_2+\delta_2(1-\lambda_2) & \alpha_2(1-\lambda_2) & \beta_2(1-\lambda_2) & \cdots & 0 & 0 \\ \lambda_3 & \delta_3(1-\lambda_3) & \alpha_3(1-\lambda_3) & \cdots & 0 & 0 \\ \lambda_4 & 0 & \delta_4(1-\lambda_4) & \cdots & 0 & 0 \\ \vdots & \vdots & \vdots & \ddots & \vdots & \vdots \\ \lambda_N & 0 & 0 & \cdots & \alpha_N(1-\lambda_N) & \beta_N(1-\lambda_N) \\ 0 & 0 & 0 & \cdots & 0 & 1 \end{pmatrix},$$

which can be compared with that given for a batch fluidized bed without baffles in Section 3.2.2. Barring return to the first cell, this process is an ordinary birth-death process with reflecting/absorbing boundary conditions. If the return probabilities, λ_i, are constant and have a rate equal to, say, λ, then the return times form a Bernoulli process with intensity λ.

We note at this point that this type of model has found a wider range of application than particle transport in continuous bubbling fluidized beds. Qi [104] formulated a very similar model for the transport of drugs in biological membranes. In addition to permeability barriers, the membranes of organism, such as bacteria, are often equipped with an active pump with a wide range of substrate specificity designed to slow the entrance of certain components such as, for instance, drugs.

To model the action of this pump in addition to the diffusive transport of drug molecules through the membrane into the organism, Qi formulated a stochastic model very similar to that given in this section. In Qi's model the jump probability to the origin models the probability of a drug molecule, in the process of diffusing into the membrane, being pumped back out by the active pump. The absorbing cell models the external of the membrane on the inside of the organism, such that a molecule entering the absorbing cell has exited the membrane and entered the organism, and can no longer enter the membrane again. A difference between Qi's model and the model above is that Qi allows for a region close to the absorbing cell where the jump-to-the-origin probability is zero.

As in process technology, modeling of such biological processes is p.t. dominated by deterministic modeling. Qi remarks in his paper that many biological processes are too complex for deterministic modeling, and that stochastic modeling deserves to be more widely explored and applied in biopharmaceutics.

5.2.4 Model Behavior and Sensitivity Analysis

In Figure 5.3 we have simulated two runs of the transport process for two different choices of the parameters. In both cases we have $N = 20$ cells, and $\delta_i = 0.2$, $\alpha_i = 0.5$ and $\beta_i = 0.3$. This results in a mean velocity inside the reactor of $v = 0.1$ and a dispersion of $2D = 0.49$ (see below). The return probabilities differ. In the left simulation they are equal to 0, and in the right equal to 0.01. The choice of 0.01 in the right simulation is somewhat, but not completely, arbitrary. We want to present an example of the Markov process that shows how the possibility of a jump back to the origin influences the evolution of the particle density and the RTD. In order to show the main effects, one has to choose the parameters in such a way that there is a reasonable but not dominating chance for a full return. In our example the speed is 0.1 cells per unit of time; hence a rough calculation shows that it takes an average $T = 200$ time units to pass through the reactor yielding an average of about two full returns.

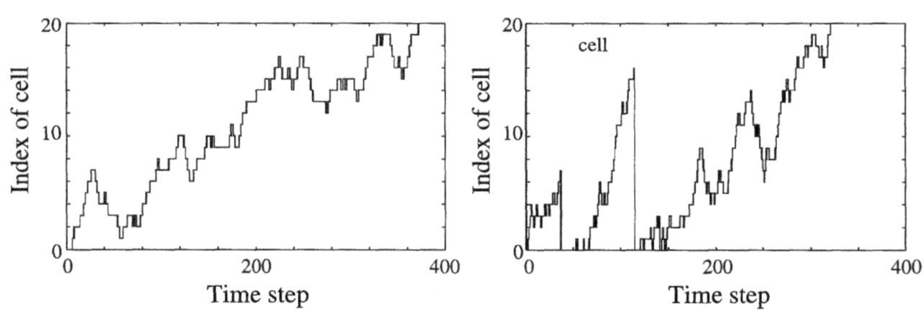

Figure 5.3: Simulation of a path in the discrete Markov model for two different parameter settings. $N = 20$ cells, $\delta_i = 0.2$, $\alpha_i = 0.5$ and $\beta_i = 0.3$ in both cases, and the return probabilities are $\lambda_i = 0$ (left) and $\lambda_i = 0.01$ (right)

In principle, one can obtain the particle distribution at any time as well as the RTD function by Monte Carlo simulations. However, this requires an extremely large number of simulations and is therefore computationally slow. For the discrete model, use of the recursion formula (5.1) is much faster. The transition matrix can be seen to be lower tridiagonal, so that (5.1) becomes:

$$p(n+1,i) = (1-\lambda_{i-1})\beta_{i-1}p(n,i-1) + (1-\lambda_i)\alpha_i p(n,i) \\ +(1-\lambda_{i+1})\delta_{i+1}p(n,i+1) \tag{5.5}$$

for $2 \leq i \leq N$, together with the boundary conditions:

$$p(n+1,1) = \sum_{i=1}^{N} \lambda_i p(n,i) + \delta_2(1-\lambda_2)p(n,2) \\ + (1-\lambda_1)(1-\beta_1)p(n,1) \tag{5.6}$$

and

$$p(n+1,N+1) = \beta_N(1-\lambda_N)p(n,N) + p(n,N+1). \tag{5.7}$$

These equations are discrete analogues of the Fokker-Planck equations.

In Figure 5.4 the probability distribution of the particle's location after $n = 25$, 50 and 100 steps for the parameter settings corresponding to the simulations in Figure 5.3 are shown. In the top row the distribution has in the interior and for small n, the characteristic Gaussian shape with center roughly at nv and standard deviation $\sqrt{2Dn}$, which for larger values of n becomes distorted by the absorbing boundary. The shape is completely different in the middle row due to the returns to the first cell. The height of the last column in these plots gives the probability that the particle has left the reactor by time n, and that is just the value of the cumulative RTD function at time n. In Figure 5.4 the complete cumulative RTD functions for the two parameter settings are also shown. The most striking difference between the two RTDs lies in the tails of the distributions. For the model with a positive return probability the tail is extremely heavy, a characteristic which is also encountered in experimental results.

In the following section we study the tail of the RTD curves making use of the tools provided by the stochastic modeling approach.

5.2.5 The Tail of the RTD Curves

Understanding the tail behavior of the RTD requires a study of the probability vector $\mathbf{p}(n) = \mathbf{p}(0)\mathbf{P}^n$ for large values of n. This can be done through

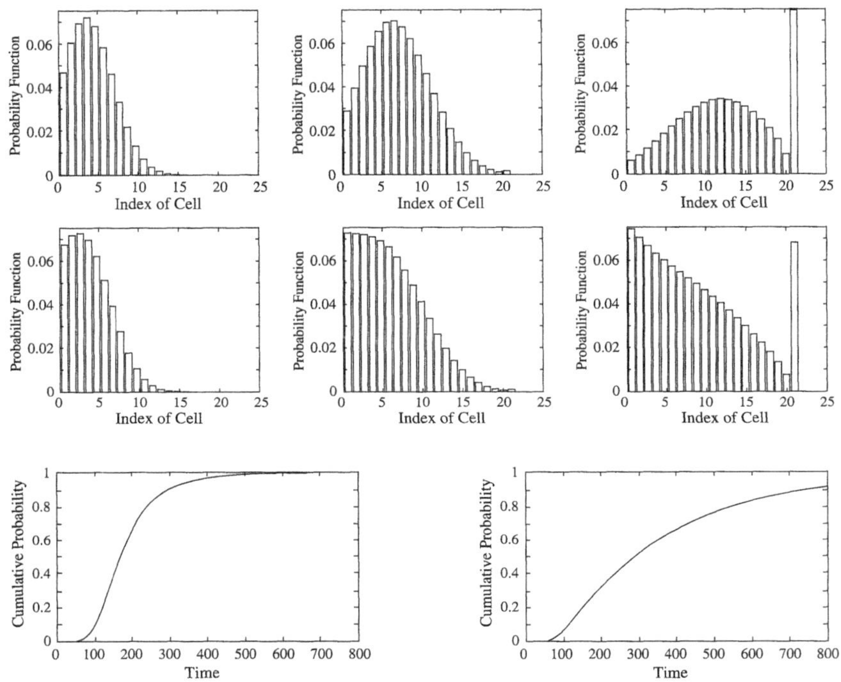

Figure 5.4: Evolution of particle density and cumulative RTD function in the discrete Markov model for two parameter settings. $N = 20$ cells, $\delta_i = 02$, $\alpha_i = 0.5$ and $\beta_i = 0.3$ in both cases, and the return probabilities are $\lambda_i = 0$ (top row and bottom left) and $\lambda_i = 0.01$ (middle row and bottom right). In the top and middle row, the density is shown at time points $n = 25$, 50 and 100 (from left to right). In the bottom row the RTD curves are shown

the eigenvalues and eigenvectors of $\mathbf{P}$. The transition matrix $\mathbf{P}$ has a unique invariant distribution, namely, $\boldsymbol{\rho} = (0, \ldots, 0, 1)$. Up to normalization, $\boldsymbol{\rho}$ is the unique left eigenvector with eigenvalue 1.

By the Perron-Frobenius theorem, all other eigenvalues are of modulus less than 1. Moreover, $\mathbf{p}(0)\mathbf{P}^n \to \boldsymbol{\rho}$ for $n \to \infty$ regardless of the initial distribution. We are interested in the difference $\mathbf{p}(n) - \boldsymbol{\rho}$, and this turns out to be governed by the second-largest eigenvalue of $\mathbf{P}$ and the corresponding eigenvector.

If $\mathbf{P}$ has a full set of left eigenvectors $\boldsymbol{\rho}, \mathbf{u}_1, \ldots, \mathbf{u}_N$ corresponding to eigenvalues $1, \mu_1, \ldots, \mu_N$ in order of decreasing magnitude, there exist constants k_1 and k_2, such that:

$$\begin{aligned} 1 - F(n) &= k_1\mu_1^n + O(\mu_2^n), \\ p(n,i) &= k_2\mu_1^n u_1(i) + O(\mu_2^n) \end{aligned} \tag{5.8}$$

for $1 \leq i \leq N$.

To show this, we can write $\mathbf{p}(0)$ as a linear combination of the eigenvectors: $\mathbf{p}(0) = a_0\boldsymbol{\rho} + a_1\mathbf{u}_1 + \cdots + a_N\mathbf{u}_N$, since the linearly independent eigenvectors constitute a base for an $N+1$ dimensional vector space. From this we obtain:

$$\begin{aligned} \mathbf{p}(n) = \mathbf{p}(0)\mathbf{P}^n &= a_0\boldsymbol{\rho} + \mu_1^n a_1\mathbf{u}_1 + \mu_2^n a_2\mathbf{u}_2 + \cdots + \mu_N^n a_N\mathbf{u}_N \\ &= \boldsymbol{\rho} + \mu_1^n a_1\mathbf{u}_1 + O(\mu_2^n) \end{aligned},$$

since $\mathbf{p}(0)\mathbf{P}^n \to \boldsymbol{\rho}$ requires $a_0 = 1$. The last coordinate in the above vector identity implies:

$$1 - F(n) = \rho(N+1) - p(n, N+1) = -a_1u_1(N+1)\mu_1^n + O(\mu_2^n) = k_1\mu_1^n + O(\mu_2^n).$$

For $1 \leq i \leq N$ we have $\rho(i) = 0$, and hence

$$p(n,i) = a_1\mu_1^n u_1(i) + O(\mu_2^n).$$

This completes the proof of Equations (5.8).

We thus see that the probability distribution inside the bed has a shape determined by $\mathbf{u}_1$. Moreover, $F(n)$ converges to 1 exponentially fast at a rate $\log \mu_1$. Hence, a fat tail of the RTD corresponds to μ_1 being close to unity. We can verify this for the second of the examples studied above. Here $\mu_1 = 0.9966$, accounting for the slow convergence of $F(n)$ to 1.

In Figure 5.5, top, we have plotted approximations $\{k_1\mu_1^n u_1(i), 1 \leq i \leq N\}$ for $n = 100$ and $n = 200$ together with the exact probability functions. For $n = 200$ the difference is no longer visible. On the bottom we have plotted the exact cumulative RTD function $F(n)$ together with $1 - k_2\mu_1^n$.

The numerical computations of, e.g., the particle distribution at time $n = 1000$ for the case of $N = 50$ cells takes less than one second on an HP workstation.

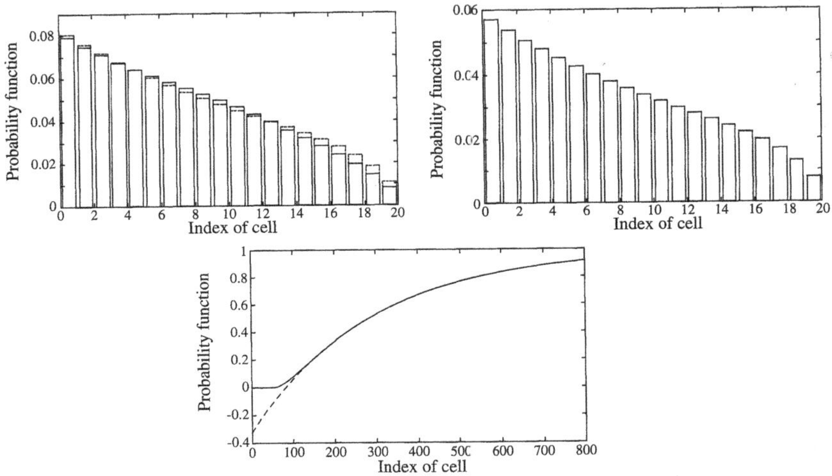

Figure 5.5: Density profiles in the interior of the bed at large times, $n = 100$ (top left), or $n = 200$ (top right), and cumulative RTD function (bottom), with parameters $N = 20$, $\delta_i = 0.2$, $\beta_i = 0.3$ and $\lambda_i = 0.01$. The dotted lines show the approximations derived from the relations in the text. Note that at $n = 200$, the approximation is almost perfect

5.2.6 Relating the Transfer Probabilities to the Physical Processes

As before for a batch fluidized bed, we denote by x the vertical distance to the top of the bed, so that $x = 0$ corresponds to the top of the bed, and $x = H$ to the bottom, where H is the total height of the bed. We divide the bed into N horizontal segments, each of height $\Delta = 1/N$. The x-coordinate of a particle in cell i is then in the range $((i-1)\Delta, i\Delta]$, $i = 1, 2, \ldots, N+1$. Time is discretized into intervals of length ε, and we will model the particle's position at times $n\varepsilon$, $n = 0, 1, \ldots$.

As for the batch fluidized bed, the parameters in the transfer probability matrix are assumed to depend on two physical parameters describing the particle transport processes:

- a superficial velocity, v, which in this case describes not only the particle "circulation" due to the transport down in the bulk compensating the upward transport in bubble wakes, which we called v_{cir}, but also an extra component, v_0 due to the particle flow through the bed, and

- a dispersion coefficient, D describing, as for batch beds, the particle dispersion caused by the stirring action of moving fluidization bubbles.

The elements of the transfer matrix are related to these parameters as follows:

$$\begin{aligned}\delta_i &= \frac{\varepsilon}{\Delta^2}D(i\Delta) - \frac{\varepsilon}{2\Delta}v(i\Delta),\\ \beta_i &= \frac{\varepsilon}{\Delta^2}D(i\Delta) + \frac{\varepsilon}{2\Delta}v(i\Delta),\\ \alpha_i &= 1 - \delta_i - \beta_i.\end{aligned} \tag{5.9}$$

This is the same as Equation (3.17), except that we, instead of v_{cir}, have written $v = v_0 + v_{cir}$. As we noted for batch fluidized beds, with these transfer probabilities, conditioned on the particle not being caught up in a bubble wake, the mean displacement per step becomes $\varepsilon v(i\Delta)$, which corresponds to a mean velocity of $v(i\Delta)$. The mean squared displacement per step is given by $\Delta^2(\beta_i + \delta_i) = 2\varepsilon D(i\Delta)$, resulting in a mean squared displacement per unit of time of $2D(i\Delta)$. D is thus the Fickian dispersion coefficient.

As an added condition, we define $D_0 = \sup_{0\leq x\leq H} D(x)$, and let ε depend on Δ via:

$$\varepsilon = \frac{\Delta^2}{4D_0}$$

This condition ensures that the probabilities α_i, β_i and δ_i are positive for sufficiently small Δ if v and D are continuous functions with values in $(0, \infty]$.

Finally λ_i, the probability that a particle is caught in a bubble wake, is, as for batch fluidized beds, calculated as:

$$\lambda_i = \frac{v_{cir}((i-1)\Delta) - v_{cir}(i\Delta)}{\Delta}\varepsilon. \tag{5.10}$$

For cell N we choose for the case of continuous beds special transfer probabilities, consistent with the physics of formation of fluidization bubbles with their associated wakes giving rise to a discrete return probability in this cell. We say that a fraction Λ of the particles arriving at the Nth cell are returned to the bed surface, and that a fraction of $1 - \Lambda$ leaves the bed to the absorbing cell, resulting in the transfer probabilities:

$$p_{N,1} = \Lambda, \qquad p_{N,N+1} = 1 - \Lambda. \tag{5.11}$$

5.2.7 The General Conditions for Danckwerts' Law to Hold for Continuous Fluidized Beds

For the fluidized bed transport process, where the transfer probabilities are given by Equations (5.3) and (5.4), the general conditions 1 and 3 in Section 4.3 for Danckwerts law for mean residence time to be satisfied become:

$$1 - \beta_1(1-\lambda_1) + \delta_2(1-\lambda_2) + \sum_{i=2}^{N} \lambda_i = 1 - v_0 \tag{5.12}$$

$$\beta_N(1-\lambda_N) = v_0 \tag{5.13}$$

Condition 2 becomes

$$\alpha_i(1-\lambda_i) + \beta_{i-1}(1-\lambda_{i-1}) + \delta_{i+1}(1-\lambda_{i+1}) = 1 \tag{5.14}$$

for $2 \le i \le N-1$ and for $i = N$

$$\alpha_N(1-\lambda_N) + \beta_{N-1}(1-\lambda_{N-1}) = 1. \tag{5.15}$$

These conditions can be simplified further by using the fact that $\alpha_i(1-\lambda_i) + \beta_i(1-\lambda_i) + \delta_i(1-\lambda_i) + \lambda_i = 1$ for $2 \le i \le N$, and $1 - \beta_1(1-\lambda_1) + \beta_1(1-\lambda_1) = 1$. Thus condition (5.14) is equivalent to

$$\beta_{i-1}(1-\lambda_{i-1}) - \delta_i(1-\lambda_i) = \lambda_i + (\beta_i(1-\lambda_i) - \delta_{i+1}(1-\lambda_{i+1})) \tag{5.16}$$

for $2 \le i \le N-1$ and

$$\beta_{N-1}(1-\lambda_{N-1}) - \delta_N(1-\lambda_N) = \lambda_N + v. \tag{5.17}$$

Iterating (5.16) and using (5.17) we thus attain

$$\beta_{i-1}(1-\lambda_{i-1}) - \delta_i(1-\lambda_i) = v + \sum_{j=i}^{N} \lambda_j, \tag{5.18}$$

for $2 \le i \le N$. Note that (5.12) yields

$$\beta_1(1-\lambda_1) - \delta_2(1-\lambda_2) = v + \sum_{j=2}^{N} \lambda_j,$$

and thus (5.18) holds for all $1 \leq i \leq N$. Together with $\beta_N(1-\lambda_N) = v$ it is actually equivalent with conditions 1, 2 and 3. Condition (5.18) is again easily interpreted: $\beta_{i-1}(1-\lambda_{i-1}) - \delta_i(1-\lambda_i)$ equals the net flow in the bulk phase through the membrane separating cell $i-1$ and cell i. This flow has to be equal to the sum of the outflow of particles and the wake-flow, at this point.

5.2.8 Comparing Model and Experiment

In Sections 3.2.3 and 3.2.5 we have given an account of how the model parameters v_{cir}, D and the removal probability λ_i are related to the physical processes occurring in a bubbling fluidized bed. It remains only to relate v_0 and Λ to the physical processes.

v_0 is obviously the volumetric in- and outflow per unit bed area.

To find Λ, we consider that for the Nth cell, the flow in from the top is $v_{cir}(N) + v_0$, and out to the absorbing cell it is v_0, so that the transfer probability to the top of the bed from that cell becomes:

$$\Lambda = \frac{(v_{cir}(N) + v_0) - v_0}{\Delta}\varepsilon = \frac{v_{cir}}{\Delta}\varepsilon.$$

Table 5.1 shows a series of cases for which the model was evaluated.

Table 5.1: Cases for which the model was evaluated

Case	v_0	$U - U_{mf}$ (m/s)	H (m)
1	4.00×10^{-3}	1.50×10^{-2}	0.10
2	4.00×10^{-3}	3.00×10^{-2}	0.10
3	2.00×10^{-3}	3.00×10^{-2}	0.10
4	5.19×10^{-5}	5.40×10^{-3}	0.0977
5	8.88×10^{-5}	2.58×10^{-3}	0.0977
6	5.26×10^{-5}	4.44×10^{-2}	0.107
7	6.78×10^{-5}	6.90×10^{-3}	0.0933
8	4.29×10^{-4}	1.55×10^{-2}	0.23
9	4.29×10^{-4}	2.35×10^{-2}	0.23

The first three cases are hypothetical, chosen to test the qualitative variation of the predicted RTD curves with the operational parameters. The last six are cases for which experimental data are available in the literature: cases 4–7 in la Rivière et al. [81], and Cases 8–9 in Morris et al. [94].

Werther [134] and others have shown that the two-phase theory generally overestimates the gas flow in the bubble phase. In the cases studied here, the fluidization velocity, U, was mostly very close to the minimum fluidization velocity, U_{mf}, and this effect is therefore not negligible. From the graphics given in the paper of Werther it was estimated that the value of $(U - U_{mf})$ should be reduced by a factor of 2/3 for cases 4–7, and by a factor of 3/4 for cases 8 and 9.

Figure 5.6 shows the results of simulating cases 1–3. The figure shows that the model predicts influences of the operating variables qualitatively in line with the experimental evidence in the literature. Note that in this and the following figures the "time" axes are dimensionless time, which is time divided by the mean residence time in the bed.

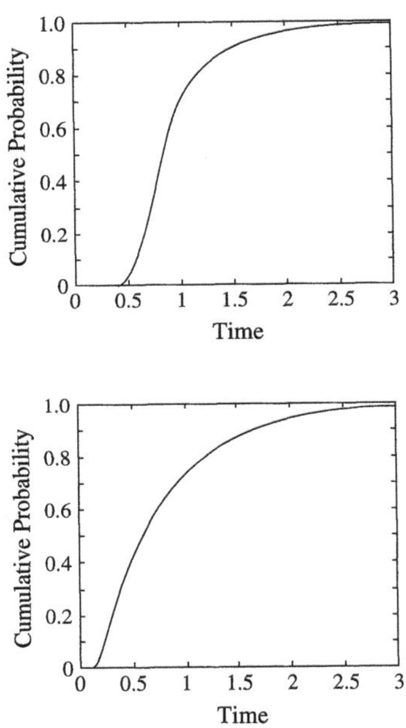

Figure 5.6: Cumulative RTD function for Cases 1–3 in Table 5.1, showing the effects of the through-flow, v_0, and the bubbling intensity, $(U - U_{mf})$

An increase in the excess superficial gas velocity, $(U - U_{mf})$, causes an increase in the mixing for equal solid through-flow, while an increase in solid through-flow for equal excess gas flow causes a tendency toward plug-flow (dispersion-free flow of particles at constant speed and without the possibility of returns to the top of the reactor). For plug-flow, the RTD curve on an x-axis of dimensionless time would be a Heaviside function $F(t) = 1_{[1,\infty)}(t)$, and for ideal mixing it would be an exponential function $F(t) = 1 - e^{-t}(t > 0)$, as shown in Appendix A. Indeed one can see in Figure 5.6 a continuous transition from near plug-flow to near ideal mixing as the bubbling intensity becomes more significant relative to the solids through-flow.

In Figure 5.7 model predictions are compared with the experimentally determined RTDs for Cases 4–9. Cases 4–7 are taken from [81]. These measurements were performed in bed of an internal diameter of 8.8 cm using colored glass particles, with tracer particles slightly larger than the bed particles. Cases 8–9 are taken from [94], who used colored tracer in a bed of 23 cm diameter. All these cases fall within the range of "bubbling fluidization" to which the model is applicable. The most interesting results are obtained for relatively low "internal flow ratios" [57], i.e. when the material flow in bubble wakes is low compared to the through-flow.

The model predictions agree quite well with the experimental points. Both prediction and experiment show an almost well mixed system in Cases 6 and 9, where the value of $(U - U_{mf})$ is relatively high. Also in the cases where the system tends to exhibit more plug-flow-like behavior (Cases 4, 5, 7, and 8) the agreement is good and, in particular, the points of initial rise of the F-curve are well predicted by the model. This shows that the envisaged mixing conditions can account for the behavior seen experimentally. It also shows that a good fit is obtained, even though the quantification of the parameters is performed on basis of the literature, so that the model in this sense does not contain any adjustable parameter.

Nevertheless, a consistent discrepancy between model predictions and experiment can be seen in Figure 5.7. The experimental points generally rise more sharply than the model predictions and in one or two cases it appears that the experimental data show a longer tail than do the model predictions.

Figure 5.8 shows a so-called intensity curve, the intensity being defined as:

$$\psi(t) = \frac{dF(t)/dt}{1 - F(t)}.$$

Under conditions of steady flow, Verloop et al. [126] showed that continuous fluidized beds exhibit a characteristic shape for $\psi(t)$, the graph exhibiting a maximum for a value of the dimensionless time between 0 and 1. The present model also shows such a local maximum for Cases 8 and 9 and a weaker one for Case 5, while in the other cases a maximum is not visible on the plots. It can also be seen in Figure 5.8 that the intensity approaches a constant value at larger values of the dimensionless residence time. This is

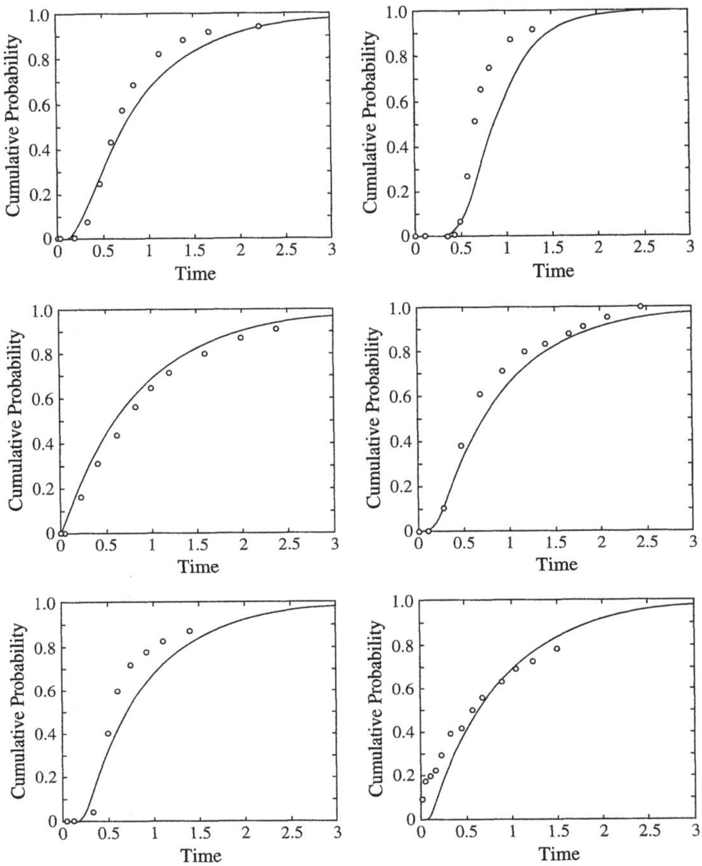

Figure 5.7: Cumulative RTD function for Cases 4–9 in Table 5.1. Points are experimentally determined data from the literature, and curves are the predictions of the model

in line with the earlier findings about the dominance of a single eigenvalue in the tail of the RTD curve.

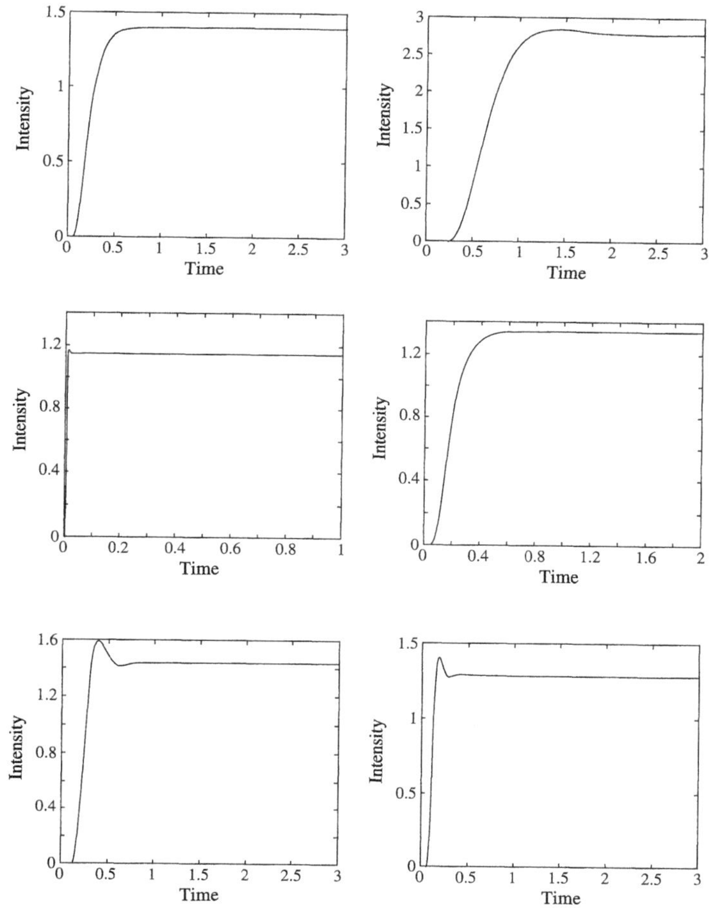

Figure 5.8: Intensity curves for Cases 4–9 in Table 5.1. The curves for Cases 6, 8, and 9 exhibit a local maximum. Such a maximum was found experimentally by Verloop et al. [126] to be characteristic of the particle RTD in continuous fluidized beds.

5.3 A Stochastic Model for Particle Transport a Fluidized Bed Riser

We now give an account of a stochastic model for a fluidized bed riser—the second configuration depicted in Fig. 5.1—proposed in a recent paper by

Harris et al. [56].

5.3.1 Physical Phenomena

Fluidized bed risers are beds operated in the transport regime, as indicated in Figure 5.1. Riser reactors, or circulating fluidized beds (CFBs) are used, for instance for the combustion of coal or for fast reactions involving the contact of gaseous reactants with a solid catalyst, such as in the fluidized bed catalytic cracking (FCC) process. In fact, in the FCC process, the catalyst circulated between two fluidized beds, one (the reactor) operated as a riser and the other (the catalyst regenerator) operated as a bubbling/turbulent fluidized bed.

For such processes, the particle residence time distribution in the fluidized bed is crucial, but is notoriously difficult to model. This has been an important impediment for the further development of this advanced type of reactor also for other applications [131]. Harris et al. [55] give a thorough discussion of these issues, and the development of CFB reactors in general.

Following Yang [138] Harris et al. [56] operate with three sub-flowregimes for fluidized bed risers: dilute phase pneumatic transport, fast fluidized bed and dense phase transport. Their model is concerned with the second of these three. The axial profile of voidage, $\epsilon = 1 - \bar{\epsilon}_s$ where $\bar{\epsilon}_s$ is the cross-sectionally averaged volume fraction taken up by the solids, is, for this flowregime, given in Figure 5.9.

Particle transport in such beds is not uniform over the cross-section. In the transition and dilute phase regions a "core-annular" flow regime exists, whereby upflow of particles occurs in a central core, which often occupies most of the cross-sectional area, and downflow takes place in an annular region on the vessel wall. An important physical parameter when modeling such a process is the thickness, δ_f of the annular region of down-flow. The correlation used for this by Harris et al [56] is one proposed by Bi et al. [13]:

$$\frac{\delta_f}{D_v} = 0.5\left[1 - \sqrt{1.34 - 1.30(\bar{\epsilon}_s)^{0.2} + (\bar{\epsilon}_s)^{1.4}}\right] \text{ for } 0.0015 \leq \bar{\epsilon}_s \leq 0.20 \quad (5.19)$$

-this corrects the equation given in [56]. D_v is the vessel diameter.

Another crucial parameter is the split of material reaching the top of the riser through the core between the fraction actually exiting the reactor, and the fraction refluxing through the annular region. van der Meer et al. [124] define the "reflux ratio" as:

$$k_m = \frac{\text{material flow refluxing through the annular film}}{\text{material flow exiting the reactor}},$$

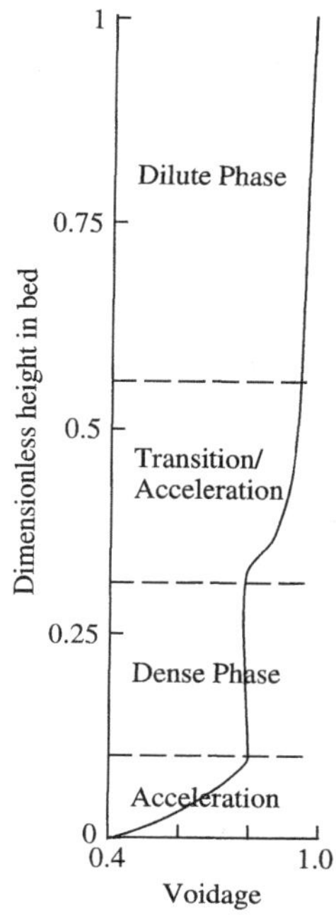

Figure 5.9: Axial profile of voidage fraction in a fast fluidized bed riser according to Yang [138] and Harris et al. [56].

and give empirical values for k_m for various configurations for the riser top, ranging from 0.11 for a short, but smooth, exit section to 2.72 for a larger, abrupt exit.

There is material exchange between the two countercurrent material flows in the core and in the annulus. According to Harris et al., very little information is available in the literature for its quantification, and they varied the extent of this interchange in their paper. We glean from their paper that they fixed the annulus→core material flow as 0.6 of the core→annulus flow for all rates of material exchange between the two regions.

5.3.2 Model

Harris et al. present four models in their paper: continuous-time and discrete versions of two different compartment configurations. The models build on the work of Fan et al. [44], which we discussed in detail in Chapter 4, and essentially consist of a number of CSTR compartments with material exchange. The two configurations are:

- A "core-annulus configuration", wherein the core and the annular regions are each split in a number of compartments axially, whereby core and annular compartments at the same elevation are allowed to exchange material in addition to the material flow up through the core compartments and down through the annular ones. The entry and exit regions constitute two additional transient compartments, and the last compartment is an absorbing one at the top.

- A "four-zone" configuration, which we will discuss below.

The former of these two is not discussed extensively by the authors, one reason being that the model predictions strongly depend on the number of axial compartments, a parameter that cannot, at the present time, be estimated from experiments with any degree of confidence. Furthermore, the continuous-time models are not discussed much in the paper either, demanding as they do Monte Carlo type simulations for their evaluation. So we will concentrate this discussion of the paper of Harris et al. on the discrete "four-zone" model.

A scheme of the "four-zone" model is shown in Figure 5.10.

The compartment labelled CT is a "churn-turbulent" (turbulent fluidization), dense-phase region at the bottom of the bed. Its height is estimated from the pressure profile over the reactor. The compartments TZ,C and TZ,A are the core and annular regions of the transition region between dense-phase and dilute regions indicated in Figure 5.9. Compartments C and A are the core and annular regions of the dilute phase region, and E is an exit region, with SRC the absorbing compartment.

The position of one solid particle is denoted by a stochastic variable $(X_t)_{t\geq 0}$, with state space $\{1, 2, \ldots, 7\}$, which is allowed to go through a Markovian process. Time is, as mentioned, discretized, so that we obtain a Markov chain $(X_n)_{n\geq 0}$. The Markov chain is specified by its transition probabilities $p_{ij}(n) := P(X_{n+1} = j | X_n = i)$. The process is assumed to be homogeneous in time, so that the argument of the p_{ij} can be dropped.

5.3.3 Quantification of the Transfer Probability Matrix

Following Harris et al. [56] we will now use the method of Fan et al. [44], discussed in Section 4.4.2 to express the transfer probabilities between the compartments. We do, however, in two important aspects deviate from the description of Harris et al., as pointed out below.

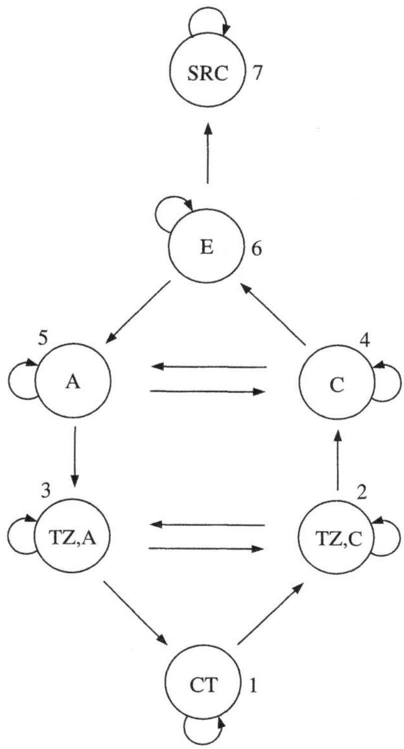

Figure 5.10: Scheme of the model of Harris et al. [56]. The CSTR compartments have been labelled as in their paper, and in addition have been numbered

Fan et al. [44] evaluated the transition probabilities from the material flows between the compartments. We repeat their equations, (4.45) and (4.46) in Chapter 4, here:

$$p_{ii} = e^{-\Delta t/\tau_i},$$

and

$$p_{ij} = \frac{Q_{ij}}{\sum_{j\neq i}^{N+1} Q_{ij}}(1 - p_{ii}).$$

In these equations, τ_i was the mean residence time in the compartment, equal to its volume divided by the volumetric through-flow

$$V_i/Q_i = V_i/\sum_{j\neq i}^{N+1} Q_{ij}$$

and Q_{ij} the flowrates from compartment i to compartment j.

In the system of Harris et al., the volume fraction taken up by solids, ϵ_s varies between the compartments, so that the material in a sense is compressible. They therefore need to consider mass flows between the compartments, m_{ij}, rather than volume flows. The equivalent to the compartment volume is then the total mass in the compartment, $M_i = \rho_p \epsilon_{s,i} V_i$, where ρ_p is the particle density. The mean residence time in compartment i, τ_i, is the total mass in the compartment divided by the total mass flow through it: $M_i/\sum_{j\neq i}^{N=7} m_{ij}$. We thus obtain the following transfer probabilities:

$$p_{ii} = e^{-\Delta t/\tau_i}, \tag{5.20}$$

and:

$$p_{ij} = \frac{m_{ij}}{\sum_{j\neq i}^{N+1} m_{ij}}(1 - p_{ii}). \tag{5.21}$$

Equations (4.45) and (4.46) deviate from those given by Harris et al. They evaluate the p_{ij} as:

$$p_{ij} = \frac{m_{ij}}{\sum_{j\neq i}^{N+1} m_{ij} + M_i} \tag{5.22}$$

(see the first equation in their Appendix A.1.), and therefore, presumably, since the p_{ij} must sum to one over j:

$$p_{ii} = \frac{M_i}{\sum_{j\neq i}^{N+1} m_{ij} + M_i} \tag{5.23}$$

The p_{ii} are therefore simply taken as the mass of solids in compartment i divided by the sum of the mass flows from the compartment and the mass in it. However, p_{ii} should be the probability of a particle *remaining* in the compartment during the given time step, and should therefore be a measure only of the fraction of material remaining at the end of the time step.

Another problem in the treatment of Harris et al. is that their time-steps are so large that the flows in and out of some of the compartments during one time step are more than the total of the material in the compartment, violating the restriction on the length of the time step mentioned in the first paragraph of Section 4.3.2 and rendering the whole concept of p_{ii} meaningless.

These are the reasons why we have preferred to apply the methodology of Fan et al. [44] rigorously using Equations (5.20) and (5.21) for quantifying the transfer probability matrix. To avoid the problem stated in the previous paragraph, we also need to impose a restriction on the time step such that for all i in the state space of X_n, $\sum_{j \neq i}^{N=7} m_{ij} \Delta t \leq M_i$, see also the first paragraph in section 4.3.2.

These deviations do change the model predictions quite a lot, as we shall see below, but not the basic shapes of the RTD-curves. We notice that in the comparisons with experimental RTD-data, Harris et al. fitted the predictions to experimental RTD data from the literature by "normalizing the time scale of the simulated RTD to the mean of the experimental curve" (p. 4790 in [56]). However, in spite of this, it is clear that the quantification of the model parameters needs to be rethought.

5.3.4 Model implementation

If values for the following parameters are known or assessed:

- The heights of the churn-turbulent, the transition, the dilute and the exit regions.
- The solids through-flow.
- The solids volume fraction, ϵ_s in the six transient compartments.
- The solids exchange rates, e.g. in the form of solids fluxes, between the core and annulus compartments of the transition and dilute regions.

Then all the solids inventories in the compartments, M_i, and the flows between them, m_{ij}, can be found by simple mass balances. To do this, however, the thickness, δ_f, of the annular regions in the transition and dilute regions need to be calculated iteratively.

In Figure 5.11 the model of Harris et al. is plotted calculating the p_{ij} using Equations (5.22) and (5.23) (gray points) and Equations (5.20) and (5.21) (black points) for the case given in Table 5.2 and a solids through-flow of 0.884 kg/s, which is similar to, but not identical to one of the cases analysed in Harris.

The relative values for the thicknesses of the annular regions, $\frac{\delta_f}{D_v}$ came to 0.132 and 0.060 for the transition and the dilute regions, respectively. k_m was taken as 0.11 (smooth exit), and the solids flux between the annular and core regions as 5 kg/m^2.

Table 5.2: Case for which the RTD model for CFBs was calculated in Figure 5.11

Comp.	Comp. No	Height (m)	$\epsilon_{s,i}$	M_i (kg)
CT,	1	1.25	0.25	12.15
TZ,C	2	0.625	0.12	1.58
TZ,A	3	0.625	0.12	1.33
C	4	2.5	0.01	0.75
A	5	2.5	0.0375	0.83
E	6	0.625	0.1	2.43

There is clearly a large difference between the predictions, although the basic shape of the E-curve with an initial sharp rise and a long tail is preserved.

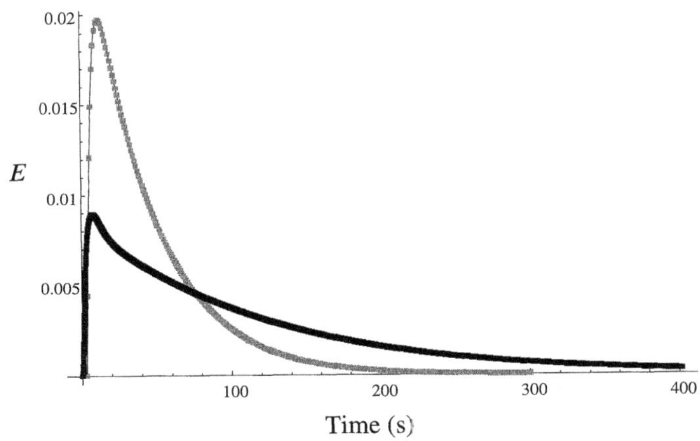

Figure 5.11: Model of Harris et al. [56] plotted calculating the p_{ij} using Equations (5.22) and (5.23) (gray points) and Equations (5.20) and (5.21) (black points) for the case outlined in Table 5.2

Figure 5.12 shows the effect of the solids flux between the annular and core regions of the transition and dilute phase sections of the riser. The effect is qualitatively the same as that shown in Figure 12 of Harris et al., namely that an increasing core/annular flux leads to a wider residence time distribution, but evaluating the p_{ij} from Equations (5.20) and (5.21) clearly makes the effect much stronger. Harris et al found a second peak developing in their E-curves at high core-annular fluxes. Such a feature is not present in Figure 5.12.

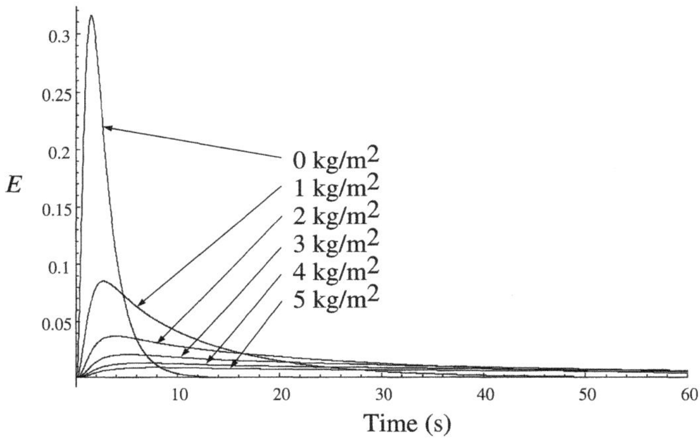

Figure 5.12: Effect of the solids flux between the core and the annulus on the particle residence time distribution according to the model of Harris et al. [56], but calculating the p_{ij} using Equations (5.20) and (5.21) for the same case as that in Figure 5.11, but varying the solids flux as indicated

We will not repeat the entire sensitivity analysis or comparison with experiment from the paper of Harris et al. Evaluating the model in accordance with the paper of Fan et al. [44], i.e. using Equations (5.20) and (5.21) will mean that the quantification of the model parameters will have to rethought. There should be plenty of room for that, since some of the model parameters, such as the core/annulus material exchange rates, are not yet quantified empirically in the literature. We think that the effect of this flux on the particle RTD shown in Figure 5.12 is physically plausible.

Chapter 6

Mixing and Chemical Reactions in Stirred Tank Reactors

In the two main sections of this chapter we will give examples of two techniques for stochastic modeling of stirred tank reactors. One is the so-called "Network-of-Zones" technique, or NoZ for short, and the other is a technique for modeling chemical reactions in continuous stirred tank reactors based on Markov chains. We shall give some results from both types of models, and show how NoZ models may be cast in the Markov-chain form to yield information complementary to the information extracted from the NoZ models in their original form.

6.1 Network-of-Zones Modeling

In this section we will discuss a special technique for modeling batch or semi-batch processes: "Network-of-Zones" modeling, or NoZ for short, introduced by professor Reginald Mann and co-workers. The systems that will be discussed are semi-batch in the sense that some components are contained as a batch within the reactor through the duration of the process, while others flow through the reactor continuously. We need, therefore, to draw on concepts for both batch and continuous processes.

In batch or semi-batch processes[1], we are often interested in finding the transient behavior of a reactor. For instance, we may suddenly add a pulse of some reactant, A, to a batch of other reactants in a reactor, and the quality of the reaction products, or the risk of developing a dangerous "hot

[1]we may loosely define a semi-batch process as a time-limited process where some components flow into and/or out of the reactor during the process, and others are contained in the reactor throughout the process

spot" due to an exothermic chemical reaction, may depend on the way A spreads through the reactor from the injection point as a function of time.

Often, however, batch or semi-batch processes may reach a steady-state concentration profile of some component. For instance in a batch fluidized bed consisting of dissimilar particles, as we considered in Chapter 3, one type of particle may tend to collect in one section of the vessel (e.g. segregate toward the bottom), and competing mechanisms for particle mixing and segregation will give rise to a steady-state concentration profile. In a semi-batch process, a reactant, which is added continuously to the reactor during the duration of the process, may be consumed due to some chemical reaction and/or flowing out of the reactor continuously. Such a reactant may exhibit at least a pseudo-steady-state concentration profile during the process period.

Stochastic modeling is eminently suited for both types of analysis: to determine the transient spatial distribution of some physical quantity, for instance the concentration of a chemical component, or to determine a final, steady-state concentration profile.

The NoZ method is the most common technique for stochastic modeling of batch or semi-batch reactors in process technology in the literature. We will discuss the method using an example that reveals the various aspects of the method, namely the NoZ model for an aerated stirred reactor by Zahradník et al. [139].

The notational conventions, the *parlange* and the solution methodology in the NoZ literature is quite different from that used in this book and in most other stochastic modeling literature, and for ease of comparison with the original literature we will be consistent with the NoZ notational conventions when giving an account of the original modeling technique.

After having discussed the original technique we then show how this type of model may be cast in the, somewhat more economical, form used in this book and how the unsteady mixing in the tank may be solved using matrix operations, rather than the CFD-like trial-and-error method used by Zahradník et al. and in other NoZ literature to find final, steady-state concentration profiles in semi-batch systems.

The NoZ method is closely related to, but simpler than, CFD modeling. Some recent work focuses on using results from CFD simulations of simplified systems as input to NoZ models that are then used to model the process in all its complexity. We discuss this briefly at the end of this section.

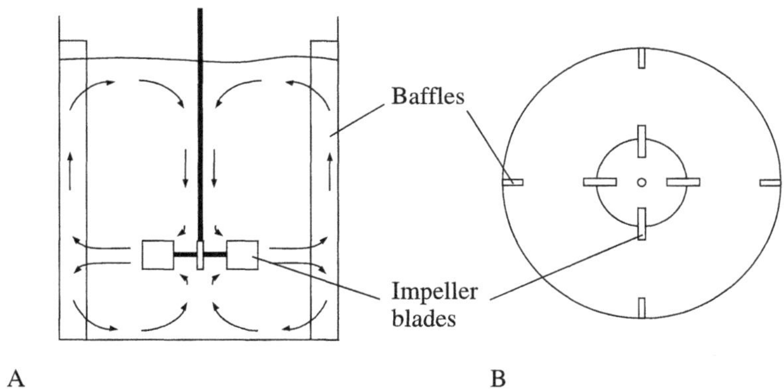

Figure 6.1: Sketch of a typical stirred tank. A: side view, the flowpattern in the liquid caused by the stirrer is indicated. B: top view

6.1.1 Stirred Tanks

Physical Phenomena

The stirred tank reactor is a system used universally in the processing industry, and therefore an appreciable amount of research, both experimental, numerical and theoretical, has been carried on this type of system.

The stirring is brought about by an impeller, most often placed on the axis of a cylindrical tank, see Figure 6.1. Vertical baffles on the wall of the tank prevent the liquid rotating as a whole with the stirrer, ensuring that it is efficiently mixed.

The effect of the rotation of a flat-bladed, and therefore non-pumping impeller is to sling out the liquid in the radial direction, giving rise to the over-all flowpattern indicated in the figure. The key to quantifying this flowpattern is to estimate the flow, q_L emanating from the perimeter of the impeller. This flow is:

$$q_L = \pi D_s W v_r \tag{6.1}$$

where D_s and W are the impeller diameter and blade width, respectively, and v_r is the radial component of the velocity with which the liquid elements leave the impeller blades.

Estimating v_r is a typical exercise in semi-empirical engineering estimation.

Figure 6.2 shows the situation: liquid elements are slung out from the blades with a velocity $\mathbf{v}$, with v_r and v_θ as the radial and tangential com-

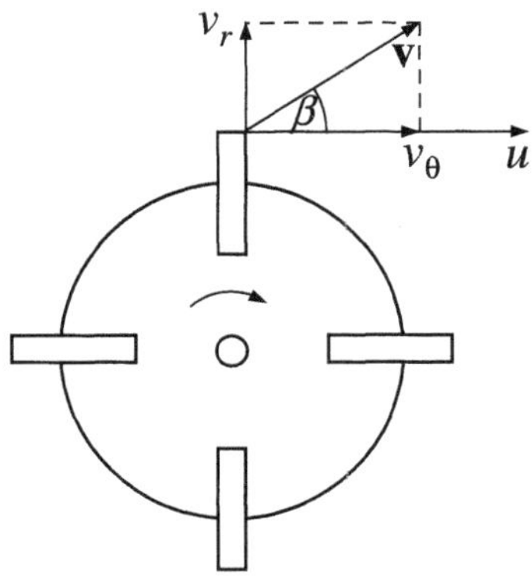

Figure 6.2: Sketch showing the velocity, $\mathbf{v}$, of liquid elements leaving the impeller blade tip

ponents, from the blades. The tips of the blades are rotating with a purely tangential velocity with tangential component u.

To estimate v_r it is generally assumed that for a given impeller the ratio v_θ/u and the angle β are constant independent of the rotational speed, such that $v_r \propto u = \pi N D_s$ where N is the rotational speed in radians per second. Furthermore, for a given impeller geometry $W \propto D_s$, and so Equation (6.1) results in:

$$q_L \propto N D_s^3 \text{ or: } q_L = K N D_s^3 \tag{6.2}$$

where the constant K takes into account all the approximations made above, and also the fact that the flow emanating from the impeller perimeter does not have a constant velocity, and that it entrains liquid from around as it emanates from the impeller perimeter. K is an impeller-specific dimensionless constant, and is often called the *flow number.*

In what follows we shall be discussing models for stirred tank reactors with three impellers. These reactors will have gas bubbles introduced into the tank at the bottom, which will flow to the surface, freeing, as they flow, oxygen necessary for the biochemical reactions in the vessel. The processes are therefore semi-batch rather than pure batch, since some components are added continuously, and others are not.

Obviously the movement of such bubbles will be influenced by the above-mentioned liquid flow pattern generated by the stirrer, but there will be a superimposed convective rise velocity driven by the boyancy force and limited by the flow force acting on the bubbles due to their rise.

Another, simpler, form of reactor, which is studied in the paper of Zahradník et al., but which we will not mention here, is an air-lift reactor, where no active stirring is present.

Original Network-of-Zones Model

Zahradník et al. [139] are ultimately interested in the concentration profile of oxygen, oxygen being a prerequisite for the biochemical reaction to proceed, in the liquid phase. As mentioned, the oxygen is continuously added to the liquid by injecting gas low in the reactor and allowing the bubbles to rise through the reactor while oxygen dissolves from the bubble gas into the liquid. Oxygen is removed from the system by escape of residual oxygen in the bubbles leaving the liquid at the surface, and by consumption due to the biochemical reaction.

The first step is to model the concentration of each size-class of bubbles in the tank and the second to model the oxygen concentration, the latter involving modeling

- the local rate of oxygen transfer into the liquid from the bubbles,
- the transport of oxygen through the liquid phase, and
- the local rate of oxygen consumption from the liquid due to biochemical reaction.

See Figure 6.3).

Note that the object is to model *steady-state concentrations* for a system where bubbles are continuously injected and oxygen is continuously consumed by reaction during the process period.

Figure 6.4 shows the general scheme of the model for the distribution of gas bubbles. The gray arrows in the figure denote convective flow due to boyancy, and the black arrows convective flow due to the over-all circulatory flowpattern in the vessel, similar to that indicated in Figure 6.2.

The system is assumed to be axisymmetrical, and the zones are toroidal in shape.

The gas introduction at the bottom is not shown, the gas bubbles exit through the surface of the liquid as indicated. The other components in the biochemical reaction system are present as batch throughout the process.

The scheme shown in Figure 6.4 can be seen to be qualitatively consistent with the flowpattern indicated in Figure 6.2. The convective flowpattern of the liquid associated with each impeller is more precisely assumed

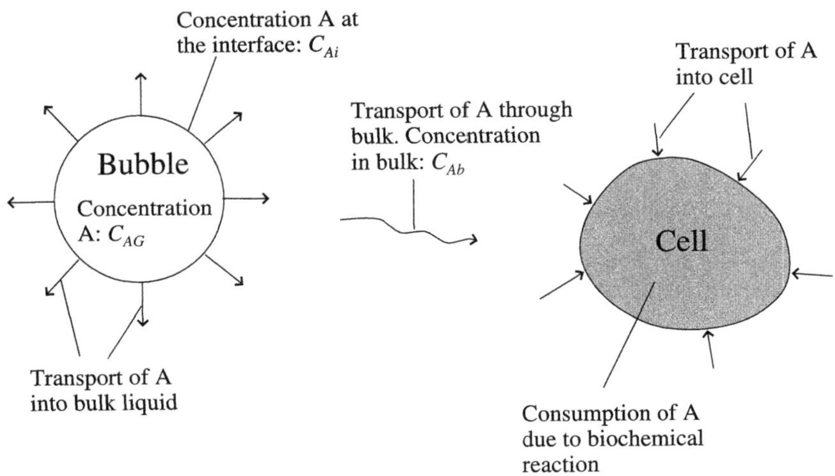

Figure 6.3: The oxygen (material A) transport processes taking place in the biochemical reactor

to consist of n separate flow-loops ($n/2$ each above and below the impeller), where n is half the number of zone-planes between the impellers (see the figure). The flow in each of these loops is assumed to be $q_L = KND_s^3/n$, such that the flow associated with each impeller sums to KND_s^3.

In addition to boyancy and loop-flows, which are convective, one additional exchange mechanism for gas bubbles between zones was considered by Zahradník et al. [139], namely turbulent dispersion, but only in the direction lateral to the direction of the impeller-generated flow (in this case in the axial direction). The total exchange scheme for one type of cell, namely one in the outward going jet from the impeller with left-to-right loop-flow, is shown in Figure 6.5.

A total balance for gas bubbles of the size class k for a zone (i, j) in the jet emanating from an impeller i.e. the type of cell depicted in Figure 6.5, then becomes:

$$\begin{aligned} &u_{bk} A \epsilon_{Gk}(i, j-1) + q_L \epsilon_{Gk}(i-1, j) \\ &+ q_L \beta \epsilon_{Gk}(i, j-1) + q_L \beta \epsilon_{Gk}(i, j+1) \\ &\qquad - q_L \epsilon_{Gk}(i, j) - q_L 2\beta \epsilon_{Gk}(i, j) - u_{bk} A \epsilon_{Gk}(i, j) = 0 \quad (6.3) \end{aligned}$$

the terms having SI units of m^3/s. u_{bk} is the velocity due to boyancy, $\epsilon_{Gk}(i, j)$ is the volume fraction bubbles of size class k in zone (i, j), and β is

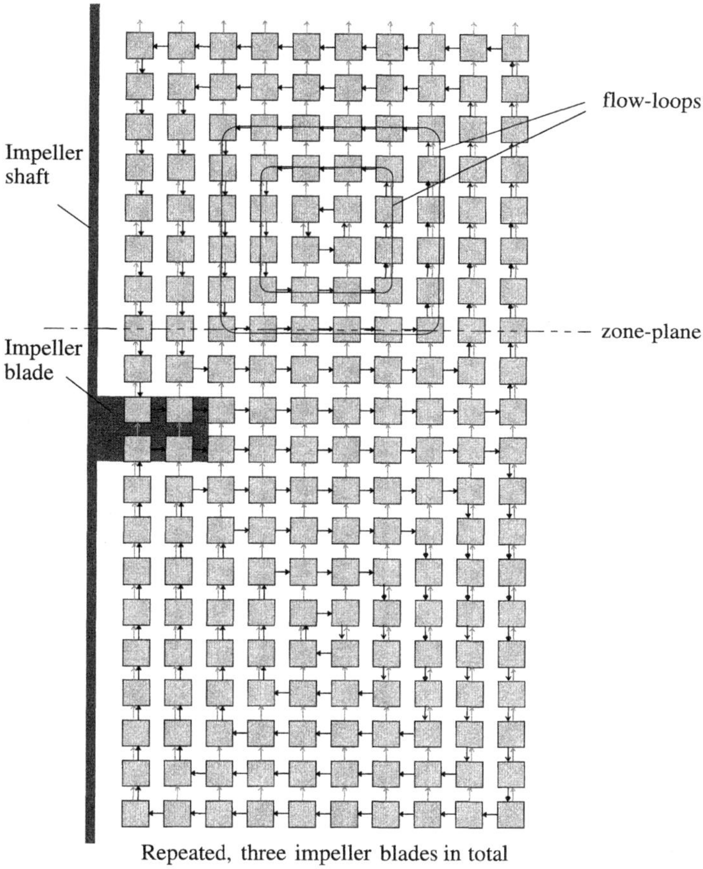

Figure 6.4: The scheme of the NoZ model of Zahradník et al. [139]. The pattern shown is repeated for three impellers in total, and the gas is injected through the central element just under the bottom impeller. Examples of zone-planes and flow-loops are indicated

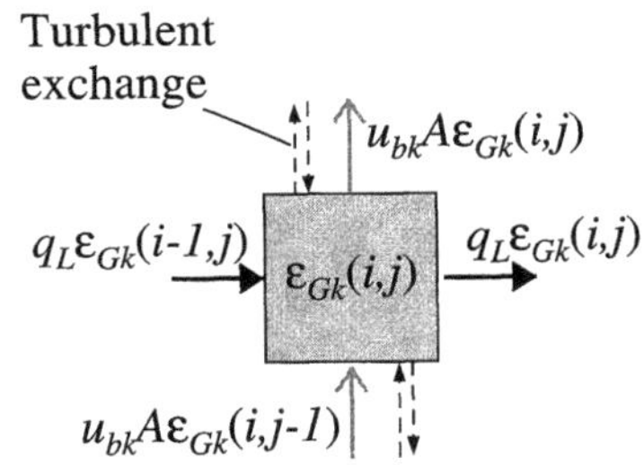

Figure 6.5: Diagram showing the exchange of gas bubbles between zone (i, j) and the surrounding zones

a dimensionless turbulent exchange coefficient to be quantified empirically. A is the horizontal "contact" area between axially neighboring zones. The terms represent in turn: inflow due to boyancy; inflow due to the liquid flowpattern generated by the impeller (loop-flow); inflow due to turbulent exchange with the zone below; inflow due to turbulent exchange with the zone above; outflow due to the liquid flowpattern generated by the impeller; outflow due to turbulent exchange with the zones above and below; outflow due to boyancy. The balance equations for zones in other positions in the flow loops are very similar.

It is stated in the paper of Zahradni'ik et al. that all of the toroidal zones have the same volume, which means that they—contrary to the impression given by Figure 6.4—become thinner at larger radial position, and that A is constant and independent of j.

Note the absence of an accumulation term in Equation (6.3), since the balance is for steady-state concentrations.

Finding the oxygen concentration in the liquid phase, C_{Ab} measured in mol/m^3, involves knowing the local rate of transfer from the bubble phase to the liquid (see Figure 6.3), which again depends on the concentration difference between the bubble gas and the bulk liquid (see Figure 6.3). Normally it is assumed that the transfer is quicker on the gas side, and therefore limited on the liquid side of the interface. The diffusional transfer process on the liquid side is driven by the difference between the interface concentration, C_{Ai} and the concentration in the bulk liquid, C_{Ab}, and is mostly assumed to be proportional to it, so that the rate of transfer per unit interfacial area is written: $k_L(C_{Ai} - C_{Ab})$ with k_L a "mass transfer coefficient" in with SI units m/s. k_L is determined empirically or, for some very simple cases, analytically.

C_{Ai}, the concentration at the interface on the liquid side, is in equilibrium with the gas phase concentration, C_{AG}, which here is measured in volume fraction or volume %. At least for low concentrations, the equilibrium concentration in the liquid is normally assumed to be proportional to the partial pressure in the gas phase: $C_{AG}p$, where p is the total pressure in the gas phase. This proportionality assumption is called Henry's law, and the proportionality constant is called Henry's constant, H^*. Thus, for bubbles of size-class k, $C_{Ai} = C_{AGk}p/H^*$ where H^* has SI units bar m^3/mol.

We now have the necessary information to write down the balance over a zone for the oxygen concentration:

$$\begin{aligned} & q_L C_{Ab}(i-1,j) + q_L\beta C_{Ab}(i,j-1) + q_L\beta C_{Ab}(i,j+1) \\ & - q_L C_{Ab}(i,j) - q_L 2\beta C_{Ab}(i,j) - r_1(i,j) \\ & \quad + \sum_k k_L \bar{a}_k(i,j)V(j)\left[C_{AGk}(i,j)p(j)/H^* - C_{Ab}(i,j)\right] = 0 \end{aligned} \tag{6.4}$$

the terms being in mol/s. $r_1(i,j)$ is the reaction rate in zone (i,j), $V(j)$ and $p(j)$ are the zone volume and pressure at radial position j, $V(j)$ actually being independent of j, as mentioned. $\bar{a}_k(i,j)$ is the interfacial area per unit volume in zone (i,j) for bubbles of size class k. If the bubbles are spherical and of diameter d_{bk}, this area becomes $\bar{a}_k(i,j) = 6\epsilon_{Gk}(i,j)/d_{bk}$. The terms represent in turn: inflow in the liquid phase due to the liquid flowpattern generated by the impeller; inflow due to turbulent exchange with the zone below; inflow due to turbulent exchange with the zone above; outflow due to the impeller generated liquid flowpattern; outflow due to turbulent exchange with the zones above and below; sink term due to consumption by biochemical reaction; source term due to transfer from bubble gas.

It remains to determine the oxygen concentration in the gas of bubbles in a given size class, $C_{AGk}(i,j)$. It is not possible to follow each individual bubble, but if it is assumed that all bubbles of a given size class have the same oxygen concentration in a given zone (i.e. essentially that the bubble gas is "well mixed" within a given size class), a zone balance equation for oxygen concentration in the bubble gas can be written for bubbles of size class k:

$$\begin{aligned} & \{u_{bk}A\epsilon_{Gk}(i,j-1)C_{AGk}(i,j-1) + q_L\epsilon_{Gk}(i-1,j)C_{AGk}(i-1,j) \\ & +q_L\beta\epsilon_{Gk}(i,j-1)C_{AGk}(i,j-1) + q_L\beta\epsilon_{Gk}(i,j+1)C_{AGk}(i,j+1) \\ & -q_L C_{AGk}(i,j) - q_L 2\beta C_{AGk}(i,j) - u_{bk}A\epsilon_{Gk}(i,j)C_{AGk}(i,j)\}/\tilde{v} \\ & -k_L\bar{a}_k(i,j)V(j)\left[C_{AGk}(i,j)p(j)/H^* - C_{Ab}(i,j)\right] = 0 \end{aligned} \tag{6.5}$$

the terms again being in mol/s. $\tilde{v}$ is the molar volume of oxygen. The terms divided by the molar volume of oxygen represent: flow in due to boyancy; flow in due to impeller generated flow; flow in due to turbulent exchange with zone above; flow in due to turbulent exchange with zone below; flow out due to impeller generated flow; flow out due to turbulent exchange with zone above and below; flow out due to boyancy. The last term is a sink term due to mass transfer to the liquid phase.

The rest of the paper is dedicated to quantifying the model parameters and studying some results of the simulations.

The same group has over the years published a series of papers [67, 68, 71, 85–87, 114, 128] on the NoZ modeling approach, mostly applied to semi-batch stirred tank reactors.

Relation of NoZ to Modeling Based on Markov Chains

NoZ modeling can quite easily be translated to a formulation based on Markov chains. We note, however, that some new aspects are contained in the NoZ models proposed in the literature, which we have not formulated in the form of Markov chains.

One of these aspects is that not all the components considered in the NoZ models are conserved, some are consumed or generated by chemical reactions. Moreover, as mentioned above, the goal of NoZ models have generally been to find a spatial concentration profile for a component that is continuously added to the reactor for at least part of the process period, rather than to determine the transient behavior of a pulse added to the reactor at $t = 0$. In other cases, e.g. [71], the transient concentration profile of a component added during only part of the processing period is considered.

Some other NoZ models, however, are directly amenable to formulation in a Markov chain model such as we have considered in this book till now, for instance Cui et al. [28] studied the mixing rate in a stirred tank by a simple NoZ model using the pulse injection of a "tracer".

In the example NoZ model in the previous section, the goal was to find the pseudo-steady-state concentration profiles for two components, which were continuously flowing through the reactor during the process period: one conserved quantity, namely the gas bubbles, and one that was consumed by a biochemical reaction, namely the oxygen. While the gas bubbles are conserved, at least if the dissolution of oxygen does not rob the bubble phase of significant gas volume, they are moving relative to the liquid, and therefore in a sense segregating to the top of the reactor.

We recall that the time-development of the position probability for one particle or molecule models the behavior of a large number of marked particles or molecules added as a Dirac pulse to the system.

In this section we will present a discrete Markov-chain model for the transient distribution of a pulse of bubbles added to the stirred tank of Zahradník et al. [139]. This will be done by formulating transition probabilities consistent with the inter-zonal flows of Zahradník et al. using the strategy of Fan et al. [44] discussed in Section 4.4.2.

We thus denote the position probability of one bubble by a stochastic variable $(X_n)_{n>0}$ with state space $\{1, 2, 3 \ldots N+1\}$, where N in this case is 600 (the total number of zones in Zahradník et al.'s model system), and the 601st cell is the absorbing cell at the surface of the liquid (see below). We allow X_n to go through a time-homogeneous Markov process, for which the $(N+1) \times (N+1)$ transfer probability matrix, $\mathbf{P}$ has elements p_{ij}. Our task is now to formulate expressions for p_{ij} consistent with the zone balances of Zahradník et al. given in the previous section.

Since the bubbles "segregate", the transition probabilities for bubbles need not satisfy the conditions in Section 4.3, and their mean residence time in the reactor liquid will, in general, not satisfy Danckwerts' law for mean residence time. However, the segregation (or rather rise) velocity is uniform throughout the reactor liquid, so that a scheme may be devised such that the columns of the transfer probability matrix nevertheless sum to unity for all the cells except the bottom ones.

Although Zahradník et al. claim that Equation (6.3) is "easily modified" to describe the other types of zones in the reactor, this is a truth with some modification. Apart from four different types of internal flow zones, there are zones at the right and left edges (zones at the tank center and wall), and in the corners. To make the task slightly easier, we consider turbulent dispersion in all directions, not only laterally to the direction of impeller generated flow, giving:

$$\begin{aligned}
& u_{bk} A \epsilon_{Gk}(i, j-1) + q_L \epsilon_{Gk}(i-1, j) \\
& + q_L \beta \epsilon_{Gk}(i, j-1) + q_L \beta \epsilon_{Gk}(i, j+1) \\
& + q_L \beta \epsilon_{Gk}(i-1, j) + q_L \beta \epsilon_{Gk}(i+1, j) \\
& \qquad - q_L \epsilon_{Gk}(i, j) - q_L 4\beta \epsilon_{Gk}(i, j) - u_{bk} A \epsilon_{Gk}(i, j) = 0 \qquad (6.6)
\end{aligned}$$

-instead of Equation (6.3).

To formulate transfer probabilities consistent with this, we use Equations (4.45) and (4.46):

$$p_{ii} = e^{-\Delta t/\tau_i} \quad \text{and} \quad p_{ij} = \frac{Q_{ij}}{Q_i}(1 - p_{ii}), \qquad (6.7)$$

where $Q_i = \sum_{j \neq i}^{N+1} Q_{ij}$ and τ_i is the mean residence time in the compartment, equal to its volume divided by the volumetric through-flow V_i/Q_i.

Like Zahradník et al., we thus assume that each zone (or cell) is ideally mixed.

We renumber the cells to have only one index i denoting the cell. We start the count at the bottom left corner, and count upward column by column, so that the left-bottom cell is $i = 1$, the bottom cell of the second column is 61, the top-right cell is cell 600, and we have one absorbing cell at the liquid surface, cell 601. Thus we may define a position probability vector $\mathbf{p}(n)$ with elements $(p_i(n))_{i=1,\ldots,601}$

For an internal cell in the jet emanating from an impeller, such as is described by the balance equation (6.6), the conditional transfer probability of a bubble to the neighboring cell to the right is:

$$p_{i,i+60} = \frac{q_L + q_L\beta}{Q_i}(1 - p_{ii}).$$

Similarly for the neighboring cells to the left, above and below, respectively, the conditional transfer probabilities are:

$$p_{i,i-60} = \frac{q_L\beta}{Q_i}(1-p_{ii}),\ p_{i,i+1} = \frac{u_{bk}A + q_L\beta}{Q_i}(1-p_{ii}),\ p_{i,i-1} = \frac{q_L\beta}{Q_i}(1-p_{ii}).$$

All other transfer probabilities from such a cell type are zero.

Note that we are only interested in the transfer probabilities *from* the cell and it is the transfer probability *conditional on the bubble being in that cell.*

Similarly we formulate conditional transfer probabilities from the other types of cell in the system. These are given in Table 6.1.

For the cell in the top left corner, the transfer probability to the cell underneath, $p_{60,59}$ is simply taken as equal to $p_{59,58}$, the transfer probability to the neighboring cell to the right $p_{60,120}$ as equal to $p_{120,180}$, and the transfer probability to the absorbing cell, $p_{60,601}$ is taken as equal to $p_{120,601}$. The probability of remaining in the top left corner cell is calculated as $p_{60,60} = 1 - \sum_{j \neq i} p_{60,j}$. This latter probability is not consistent with the right-hand equation of (6.7), but could be made consistent with it by adjusting the volume of the cell. Similarly for the right upper corner cell: $p_{600,599} = p_{599,598}$, $p_{600,540} = p_{540,480}$, $p_{600,601} = p_{540,601}$, and the probability of remaining in the cell, $p_{600,600} = 1 - \sum_{j \neq i} p_{600,j}$, which again can be made consistent with (6.7) by volume adjustment.

No extensive strategy for assigning the probabilities to the bottom cells is embarked upon, the columns of the transfer probability matrix do not sum to 1 for these cells. Since the flow of bubbles is upward seeking, only few bubbles arrive at the bottom cells.

All other transfer probabilities were set to zero.

Table 6.1: Conditional transfer probabilities for gas bubbles in the stirred tank reactor, all terms are to be multiplied by $(1 - p_{ii})$. For all cells $p_{ii} = e^{-\Delta t/\tau_i}$

Cell type	$p_{i,i+60}$	$p_{i,i-60}$	$p_{i,i+1}$	$p_{i,i-1}$	$p_{i,601}$
internal, right flow	$\frac{q_L+q_L\beta}{Q_i}$	$\frac{q_L\beta}{Q_i}$	$\frac{u_{bk}A+q_L\beta}{Q_i}$	$\frac{q_L\beta}{Q_i}$	0
internal, left flow	$\frac{q_L\beta}{Q_i}$	$\frac{q_L+q_L\beta}{Q_i}$	$\frac{u_{bk}A+q_L\beta}{Q_i}$	$\frac{q_L\beta}{Q_i}$	0
internal, upflow	$\frac{q_L\beta}{Q_i}$	$\frac{q_L\beta}{Q_i}$	$\frac{u_{bk}A+q_L+q_L\beta}{Q_i}$	$\frac{q_L\beta}{Q_i}$	0
internal, downflow	$\frac{q_L\beta}{Q_i}$	$\frac{q_L\beta}{Q_i}$	$\frac{u_{bk}A+q_L\beta}{Q_i}$	$\frac{q_L+q_L\beta}{Q_i}$	0
left, right flow	$\frac{q_L+q_L\beta}{Q_i}$	0	$\frac{u_{bk}A+1.5q_L\beta}{Q_i}$	$\frac{1.5q_L\beta}{Q_i}$	0
right, left flow	0	$\frac{q_L+q_L\beta}{Q_i}$	$\frac{u_{bk}A+1.5q_L\beta}{Q_i}$	$\frac{1.5q_L\beta}{Q_i}$	0
left, upflow	$\frac{q_L\beta}{Q_i}$	0	$\frac{u_{bk}A+q_L+1.5q_L\beta}{Q_i}$	$\frac{1.5q_L\beta}{Q_i}$	0
right, upflow	0	$\frac{q_L\beta}{Q_i}$	$\frac{u_{bk}A+q_L+1.5q_L\beta}{Q_i}$	$\frac{1.5q_L\beta}{Q_i}$	0
left, downflow	$\frac{q_L\beta}{Q_i}$	0	$\frac{u_{bk}A+1.5q_L\beta}{Q_i}$	$\frac{q_L+1.5q_L\beta}{Q_i}$	0
right, downflow	0	$\frac{q_L\beta}{Q_i}$	$\frac{u_{bk}A+1.5q_L\beta}{Q_i}$	$\frac{q_L+1.5q_L\beta}{Q_i}$	0
top, left flow	$\frac{1.5q_L\beta}{Q_i}$	$\frac{q_L+1.5q_L\beta}{Q_i}$	0	$\frac{q_L\beta}{Q_i}$	$\frac{u_{bk}A}{Q_i}$
bottom, left flow	$\frac{1.5q_L\beta}{Q_i}$	$\frac{q_L+1.5q_L\beta}{Q_i}$	$\frac{u_{bk}A}{Q_i}$	0	0
bottom, upflow	$\frac{1.5q_L\beta}{Q_i}$	0	$\frac{q_L+1.5q_L\beta+u_{bk}A}{Q_i}$	0	0

The probability distribution $\mathbf{p}(n)$ for a bubble initially placed in some cell in the tank can now be computed by successive matrix multiplication. Either by the recursion relation:

$$\mathbf{p}(n) = \mathbf{p}(n-1)\mathbf{P},$$

or by:

$$\mathbf{p}(n) = \mathbf{p}(0)\mathbf{P}^n.$$

See also Equations (3.15) and (3.16).

Figure 6.6 shows the results of a series of simulations where $p_8(0) = 1$ and $p_i(0)_{i \neq 8} = 0$. The other parameters were quantified in the following manner:

Reactor diameter:	1 m,
Stirrer diameter, D_s:	0.333 m,
Liquid height:	2 m,
Flow number, K:	1.3,
Liquid density, ρ:	1000 kg/m^3,
Liquid viscosity, μ:	1×10^-3 kg/ms,
Bubble diameter, d_{bk}:	2 mm,
Bubble velocity, from Stokes law: $u_{bk} = (d_{bk}^2 \rho g)/(18\mu)$	2.15 m/s,
Dispersion coefficient, β:	0.2,

The cell volumes were, following Zahradník et al., taken as constant, and the time step, Δt, was 0.002 seconds for these calculations, satisfying the restriction on the length of the time step mentioned in Section 4.3.2.

The three profile plots A, B and C in Figure 6.6 show, through the development of the probability distribution for one bubble undergoing a Markov process, how a pulse of a large number of bubbles initially in cell 8 will be drawn to the first impeller in the bottom flow loop, be entrained in the jet from this impeller, and how most of the cloud, due to the upward flow driven by boyancy, will be entrained in the flow-loop above the first impeller, all the while being dispersed by turbulent dispersion.

The fourth profile plot, D, is rescaled as indicated in the figure caption to show the detail of the probability distribution at the last time step simulated. It is seen that a small part of a bubble cloud will be entrained also in the bottom flow-loop.

In this way it is possible by simple matrix operations to simulate the distribution of a component in a stirred tank. Casting the model in a Markov process form, and solving it using matrix multiplication as we have done here gives information about the transient behavior of the reactor,

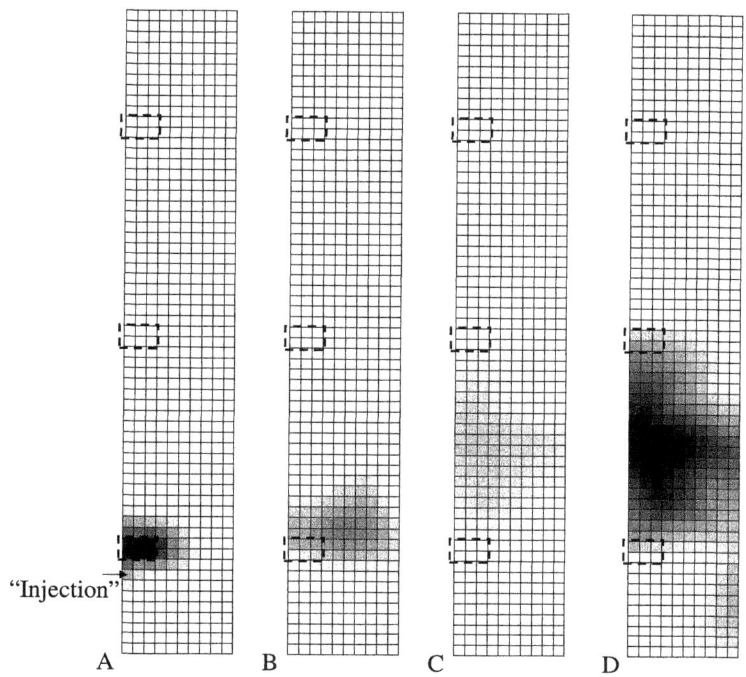

Figure 6.6: A series of profile plots showing the development in time of the probability distribution for a gas bubble placed in cell 8 at $t = 0$, and therefore, the development of a pulse of a large number of gas bubbles injected into cell 8 at $t = 0$. A: after 16 time steps, B: after 36 time steps, C: after 96 time steps. These are all plotted with white denoting zero probability and black denoting the maximum probability (0.0729) after 16 time steps. D: the profile plot in C replotted with black denoting the maximum probability (0.0105) after 96 time steps.

complementary to that of the original model of Zahradník et al. [139]. We stress that, although the transfer probability matrix in Table 6.1 looks more complicated than Zahradník et al.'s presentation of the original model, this is only because we have given the full transfer probability matrix here, while Zahradník et al. chose to display only one example balance equation, not giving the equations for the other three types of zones, or for zones at the edges and the corners of the domain.

We note, as is discussed elsewhere in this book, that also the invariant, or steady-state, concentration profile can be found from the Markovian analysis.

6.1.2 Zone Generation and Inter-Zonal Interchange Rates for NoZ Based on CFD

We briefly describe one of the latest developments in NoZ modeling. As in other stochastic process models, input is required to quantify the inter-zonal flows in NoZ models. In, for example, the model above for a tank reactor with stirrers and the fluidized bed models described elsewhere in this book this input is provided by empirical process information.

Another strategy, proposed by Bezzo et al [10, 11], is to obtain the necessary input from CFD. This would only be possible for some processes. For example, it would be possible for processes in stirred tanks, such as the one described above, where the flow pattern of the liquid can be simulated reasonably well with CFD, but it would not be possible for e.g. fluidized bed processes, where the transport of the particles cannot be quantified using CFD, at least at present and for the foreseeable future.

In principle *all* the information for the process described in Section 6.1.1 could be obtained from CFD and the NoZ approach can, in fact, be seen as a simplified form of CFD. The advantage of using the NoZ approach even for processes that in principle are amenable to CFD simulation, is that it is far more computationally efficient than CFD, and can therefore simulate processes that are so complex that their simulation by CFD is not practicable.

For this reason, the NoZ approach has been most successful for complex processes, such as biochemical and crystallization processes.

We will not give a detailed account of the work of Bezzo et al. here, but only introduce the concepts.

In the first of two papers [11], the authors divide the reactor under investigation into a number of zones, as done in the example in Section 6.1.1. In these zones the relevant processes are modeled and between there is material interchange between the zones. In the above example each of the zones were well mixed, biochemical processes took place consuming some

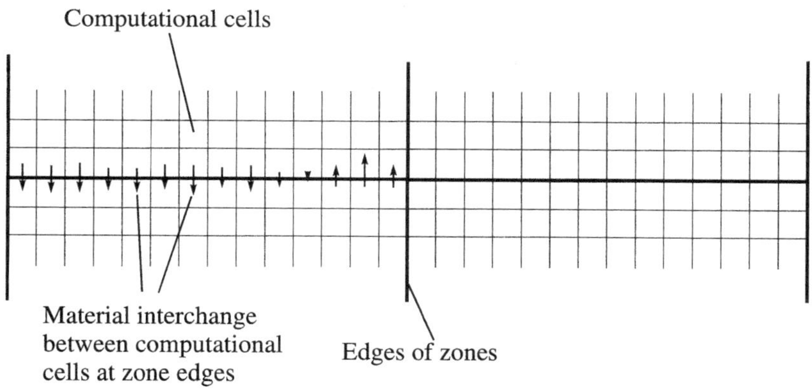

Figure 6.7: Sketch illustrating the principle of estimating material exchange between zones from CFD by summation

of the components, and material exchange took place due to the liquid flowpattern, the turbulent exchange and the bubble flow due to boyancy.

Bezzo et al. [11] propose to perform CFD simulations on the reactor under investigation to quantify the zonal material exchange rates rather than quantifying these rates from models, such as the loop-flow model in the previous sections. Computational cells for CFD will normally be much smaller than the zones for NoZ modeling, and in a mathematical formulation, Bezzo et al. propose to let each zone comprise a subset of the computational cells. The material exchanges between two zones can then be computed by simple summation over the computational cells lining the border between them (see Figure 6.7), with the possibility that both forward and backward flows may be non-zero, since the flow direction is not necessarily the same in all the computational cells.

A problem is that the physical parameters determining the exchange rates—both convectional due the over-all flowpattern and dispersional due to turbulence—may depend on the calculations of the NoZ model. For instance, in a crystallizer the apparent density and viscosity of the suspension may depend on the concentration and size of the crystals present. An iterative scheme is therefore required, such that the two simulations may feed each other information. Information from the NoZ simulation must be distributed in some reasonable manner over the computational cells comprising a zone, for instance it may be a problem for the CFD simulation if there is a step-change in physical properties between the cells bordering two zones with different properties.

In the second of their two papers, [10] the authors propose a method for identifying regions in the CFD simulated flowpattern that are suitable for aggregation to one zone in the NoZ model, and thus automatize the zone generation based on the CFD simulations.

6.2 Stochastic Modeling of Chemical Reactions in Continuous Reactors

Chemical reactions are stochastic of nature, and a large body of research literature is dedicated to stochastic modeling of chemical reactions. This is outside the scope of this book, which is specifically dedicated to process technology.

However, continuous reactors with chemical reactions is well within the scope of process technology, and below we will give an account of a methodology for modeling such systems based on Markov chains, as always using an example.

During the 1980's Too, Fan, Nassar and co-workers published a series of articles [44, 95–98, 117–119] wherein they proposed models for a variety of reactors and reactions based on Markov chains. As an example of stochastic models for continuous reactors with chemical reactions we will discuss their paper on complex chemical reactions [117], since it involves most of the features of this type of model.

6.2.1 The Reactor and the Reactions

The models we are describing are for a continuous stirred tank reactor (CSTR), wherein a chemical reaction takes place. As for the NoZ model example discussed in Section 6.1.1, the model of Too et al. is intended to describe the distribution over the possible states of material *continuously added to the reactor* with time, and eventually a distribution which is steady-state, except for the accumulation in the absorbing cell, which constitutes the exterior of the reactor.

This section therefore presents one method of *extending stochastic modeling based on Markov chains to a process with continuous inflow of material.*

As discussed in Chapter 4, in a CSTR the composition of the effluent is the same as that in the reactor, and the mean residence time of molecules or fluid elements is equal to the reactor volume, V divided by the throughflow, which we here denote by q.

The rates of chemical reactions are expressed in moles reacted per unit volume per second and can often be described by a rate constant, k, multiplied by some power of the reactants concentrations. In this section we will discuss two of the types of reactions considered in the paper by Too et al. [117].

The first type is the reversible dissociation reaction:

$$\mathrm{A}_1 \underset{k_2}{\overset{k_1}{\rightleftharpoons}} 2\mathrm{A}_2. \tag{6.8}$$

We first consider the forward reaction. This is a unimolecular dissociation reaction, the rate of which we assume can be described by a rate constant times the number of moles[2] present in the reactor (please see below for a physical reasoning behind this). If we call the number of moles and molecules of A_1 in the reactor N_1 and n_1, respectively, we can write:

$$-\frac{dN_1}{dt} = k_1 N_1 \;\Rightarrow\; -\frac{dn_1}{dt} = k_1 n_1 \tag{6.9}$$

where we have called the rate constant k_1 and the second equation in terms of number of molecules is obtained from the first by multiplying through with Avogadro's number, Av.

The backward reaction is a bimolecular association reaction, the rate of which we assume can be written $k_2 N_2^2$, where k_2 is the reaction rate constant and N_2 is the number of moles of A_2 molecules in the reactor (see below for a physical reasoning). Using a similar nomenclature as above, we can write:

$$-\frac{dN_2}{dt} = k_2 N_2^2. \tag{6.10}$$

We cannot rewrite this rate equation from its conventional from in terms of moles to one in terms of molecules maintaining the same reaction rate constant as we could Equation (6.9). To rewrite it we need to define a new rate constant, k_2', for number of molecules as follows:

$$-\frac{dN_2}{dt} = k_2 N_2^2 \;\Rightarrow\; -\frac{dn_2}{dt} = k_2 n_2 N_2 = k_2 n_2 \frac{n_2}{Av} := k_2' n_2^2. \tag{6.11}$$

The second type of reaction scheme we will consider is the competitive-consecutive reaction:

$$\begin{aligned} A_1 + A_2 &\xrightarrow{k_1} A_3 + A_5, \\ A_1 + A_3 &\xrightarrow{k_2} A_4 + A_5, \end{aligned} \tag{6.12}$$

[2] A mole of material is Avogadro's number, Av of molecules. A mole was originally defined as 12 g of pure carbon-12, which contains Avogadro's number of carbon atoms

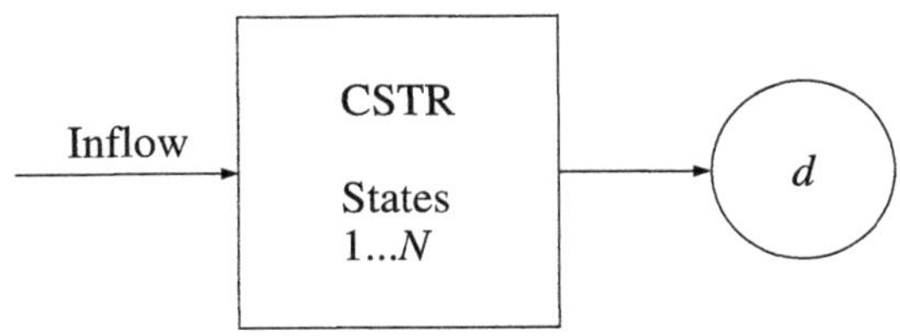

Figure 6.8: Scheme of the model for a continuous CSTR with chemical reaction. d is the absorbing state

a reaction scheme too complicated to analyze by conventional analytical modeling techniques.

6.2.2 Models

The basis of the models of Too et al. [117] for CSTRs with chemical reaction is to model a molecule's state as a stochastic variable X_m undergoing a Markovian process discrete in time and space, whereby the state space for X_m includes not only the molecule's possible spatial positions, but also its possible chemical forms. A scheme is shown in Figure 6.8, where the spatial cell comprising the reactor contains a number of different states corresponding to the number of chemical species under investigation. d is the absorbing state.

As before, the discrete Markov process is determined by the transfer probability matrix $\mathbf{P}(m)$, which may be a function of time, m, and by the initial probability vector, $\mathbf{p}(0)$.

We have several times in this book pointed to the fact that by the law of large numbers, the probability distribution of one molecule, which we focus on with process models based on Markov chains, is equivalent to the distribution over the possible states of a pulse of molecules or particles. In the models described in this section, a continuous inflow of material is dealt with as pulses of molecules added to the process each time step. The result of the Markov chain therefore looses its character of representing the probability distribution for one particle and it is easier to understand the physical meaning of the model result as the spread of material continuously added to the system.

6.2.3 Model for a CSTR with a Reversible Dissociation Reaction

In the first of the two systems discussed by Too et al., which is a continuous reactor in which the reversible dissociation reaction (6.8) is carried out, the three possible states are:

1. A_1 in the reactor,
2. a pair of A_2 in the reactor or
3. absorbed.

We require the transfer probability matrix $\mathbf{P}$ with elements p_{ij} between these three states.

Consider first the forward reaction in (6.8). We denote by $\zeta_1(m)$ the probability that a given A_1 molecule will dissociate during the m'th time interval of duration Δt. Physically this depends on the probability that it sometimes during Δt possesses sufficient energy to overcome the energy barrier between forms A_1 and $2A_2$. The rate of such a reaction in moles or molecules of A_1 dissociated in the reactor per second is thus:

$$-\frac{dn_1}{dt} = \frac{\zeta_1(m)}{\Delta t} n_1(m). \tag{6.13}$$

Comparing this with Equation (6.9), we see that we can identify $(\zeta_1(m)/\Delta t)$ with k_1.

The reverse reaction is not so easy. The ideal situation is to have a particle that can change between states, but remains identifiable. In this case, however, we have double the particles in the A_2 state, and the association of one of them to form an A_1 molecule implies the association of another. One way around this problem is to "tag" one of the A_2 molecules with the identity of the particle, and to assess its probability of transferring (possibly back) to the A_1 state. Physically the probability of any A_2 molecule associating with another will depend on the probability that

- the A_2 molecule will collide with another A_2 molecule,
- the collision is with sufficient energy and under the right angle for the association reaction to take place.

We can therefore set this probability equal to a constant times the number of other A_2 molecules the the reactor: $\zeta_2'[n_2(m) - 1] \approx \zeta_2' n_2$. The rate of the association reaction in number of *tagged* molecules per second is thus:

$$-\frac{dn_2}{dt} = \frac{\zeta_2'(m)}{\Delta t} n_2(m) \frac{1}{2} n_2(m) = \frac{\zeta_2(m)}{\Delta t} n_2^2(m). \tag{6.14}$$

Comparing with Equation (6.10), we can identify $\zeta_2(m)/\Delta t$ with k_2', and therefore with the conventional rate constant in terms of moles divided by Avogadro's number: k_2/Av.

Since the reactor is well mixed, the probability of any molecule of transferring to the absorbing state at time step m, $\mu(m)$ (the external of the reactor) is simply: $\mu(m) = q(m)/V(m)$.

We now have expressions for all the transfer probabilities, and setting the probabilities $p_{ii} = 1 - \sum_{j \neq i}^{N} p_{ij}$, where N, the number of states in this case is 3, we obtain for the total transfer probability matrix:

$$\begin{pmatrix} 1 - \zeta_1(m) - \mu(m) & \zeta_1(m) & \mu(m) \\ \zeta_2(m)n_2(m) & 1 - \zeta_2(m)n_2(m) - \mu(m) & \mu(m) \\ 0 & 0 & 1 \end{pmatrix}$$

The transfer probability matrix depends on n_2, the number of A_2 molecules in the reactor, and this therefore needs to be computed during the computational scheme. This, together with the need to add molecules from the "inlet" at each time step makes the computational scheme more complicated than just successive matrix multiplications. We adopt the following scheme:

1. Add molecules from the inlet.
2. Work out **P**.
3. Work out **p**, which is equivalent to the fraction of molecules in each state, from the total number of molecules in each state.
4. Perform the matrix multiplication $\mathbf{p}(m) = \mathbf{p}(m-1)\mathbf{P}$ to find the new distribution over the states.
5. Work out the number of molecules in each state from $\mathbf{p}(m)$.
6. Go back to 1.

Figure 6.9 shows the results of this model using the parameters:

Inflow, q:	$0.01\ \text{m}^3/\text{min}$
Reactor volume, V:	$1\ \text{m}^3$
Forward rate constant, k_1:	$0.003\ \text{min}^{-1}$
Backward rate constant, k_2:	$0.002\ \text{M}^{-1}\text{min}-1$
Initial molar conc's in reactor:	$C_{1,0} = C_{2,0} = 0\ \text{m}^{-3}$
Molar conc's in the feed:	$C_{1,f} = 1.0\ \text{m}^{-3}$ and $C_{2,f} = 0\ \text{m}^{-3}$.

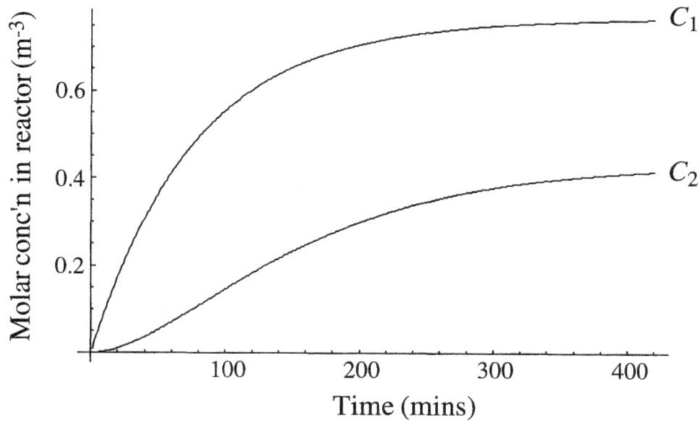

Figure 6.9: Concentrations of A_1 and A_2 reacting in a CSTR according to scheme 6.8) as functions of time. This plot is equivalent to Figure 1(a) in [117].

For easier comparison with Too et al., we have stuck to their practice of measuring time in minutes rather than seconds.

It is clear how it takes some time for A_2 to begin to form in the reactor after injection of A_1 begins. This simple reaction is also amenable to analytical solution, and Too et al. compare these curves with analytical solutions, finding excellent agreement.

6.2.4 Model for a CSTR with a Competitive-Consecutive Reaction

More interesting is the second reaction scheme given above (6.12), which is not amenable to analytical solution. The full transition matrix for this scheme is also given by Too et al. and is:

$$\begin{pmatrix} \begin{matrix}1-\zeta_1 n_2\\ -\zeta_2 n_3-\mu\end{matrix} & 0 & \frac{1}{2}\zeta_1 n_2 & \frac{1}{2}\zeta_2 n_3 & \begin{matrix}\frac{1}{2}(\zeta_1 n_2\\ +\zeta_2 n_3)\end{matrix} & \mu \\ 0 & 1-\zeta_1 n_1-\mu & \frac{1}{2}\zeta_1 n_1 & 0 & \frac{1}{2}\zeta_1 n_1 & \mu \\ 0 & 0 & 1-\zeta_2 n_1-\mu & \frac{1}{2}\zeta_2 n_1 & \frac{1}{2}\zeta_2 n_1 & \mu \\ 0 & 0 & 0 & 1-\mu & 0 & \mu \\ 0 & 0 & 0 & 0 & 1-\mu & \mu \\ 0 & 0 & 0 & 0 & 0 & 1 \end{pmatrix},$$

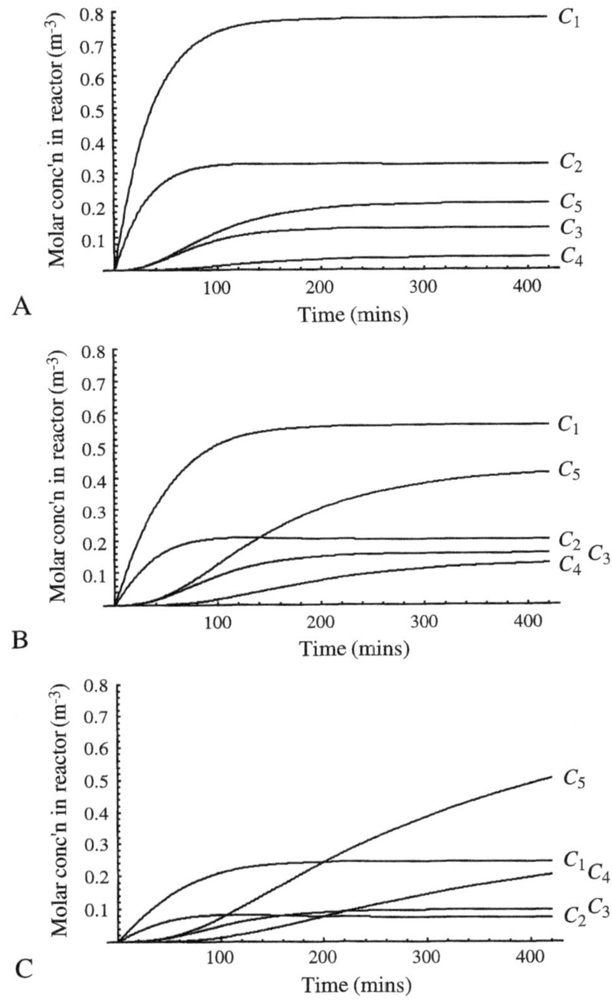

Figure 6.10: Concentrations of chemical species reacting in a CSTR according to reaction scheme 6.12) as functions of time. Three different reactor volumes are used. Plot B is equivalent to Figure 5(a) in [117].

where we remember that the number of molecules of A_1 and A_2 in the reactor, n_1 and n_2, respectively, are functions of the time step, m. Results of from this model are shown in Figure 6.10 where the following parameters were used:

Inflow, q:	0.01 m^3/min
First rate constant, k_1:	0.025 M^{-1}min-1
Second rate constant, k_2:	0.015 M^{-1}min-1
Initial molar conc's in reactor:	$C_{1,0} = C_{2,0} = C_{3,0} = C_{4,0} = C_{5,0} = 0$ m^{-3}
Molar conc's in the feed:	$C_{1,f} = 1.0$ m^{-3}, $C_{2,f} = 0.5$ m^{-3} and $C_{3,f} = C_{4,f} = C_{5,f} = 0$ m^{-3}.

Three different reactor volumes are used: $V = 0.5$ m^3, $V = 1$ m^3 and $V = 3$ m^3, in A, B and C, respectively.

The figure clearly shows how the relative concentrations and the selectivity between the products change as the volume of the reactor is varied.

In some of the other papers by Too et al. mentioned in the beginning of this section, this or similar methods are used on other types of reacting systems, for instance a polymerization reaction [96] and in a model for the gaseous reactants in a fluidized bed reactor [119], several spatial cells in the reactor are considered in addition to two different chemical states.

Chapter 7

Particle Size Manipulation

In this chapter we describe stochastic models for processes wherein the size of solid particles changes, either spontaneously or by being manipulated in some way.

By far the most used modeling technique for this type of process is Population Balance Modeling, or PBM, and we start the chapter with a discussion of PBM in general and subsequently give an example of a PBM analysis of an agglomeration process.

In the second main section of the chapter we give an account of work aimed at linking traditional PBM analysis of a grinding, also called comminution, process to a stochastic model based on discrete Markov chains.

PBM is a large and rapidly developing research field. A search in the ISI database in October of 2006 for the topic *("population balance*" and model*) and (partic* or drop* or cell*)* yielded 98 papers using PBM published just in 2006. Most of these articles were concerned with process technology applications, such as

- crystallization, where solid particles nucleate and grow,
- granulation, and other processes where solid particles agglomerate, and
- comminution, where solid particles are broken.

A significant number of articles where, however, also concerned with applications within the life sciences, for example modeling the processes of cell division and cell aggregation. Lately PBM has increasingly been extended to the study of other particle properties than size, for example particle composition or particle density. Also the particle spatial position within the process vessel has been included in the particle properties. We will discuss this below.

Scientific conferences focusing exclusively on PBM are held, and special journal issues dedicated to PBM include Chemical Engineering Science Volume 57, issue 12 and Volume 61, Issue 1.

PBM theory builds on stochastic analysis, although it most often applied in an essentially deterministic way using the law of large numbers. A fully stochastic PBM formulation for small populations is also possible.

7.1 Processes for Particle Size Manipulation, Physical Phenomena

In this section we introduce the processes of crystallization, agglomeration and comminution and mention some of the most important physical aspects and methods for carrying out the processes. More detailed information can be found in e.g. Perry's Chemical Engineering Handbook [101] or Coulsons & Richardson's Chemical Engineering [26].

7.1.1 Crystallization

In crystallization a solid particulate product is formed by allowing a solution of the material, which is to be crystallized out, to reach *supersaturation*, i.e. a concentration higher than the equilibrium solubility. Supersaturation may be brought about either by

- evaporating solvent, by adding heat and/or lowering the pressure, or
- by cooling the solvent, which for most systems will lower the equilibrium solubility.

See also Figure 7.1

The size distribution of the crystals produced, which is often an important criterion for product quality, depends on the degree of supersaturation, the degree of agitation of the solvent/crystal suspension, the rate of evaporation or cooling, and the presence or absence of nucleation sites. Nucleation sites may be provided by the asperities on rough container walls or artificially by particles added to the process vessel, either inert particles or crystallites of the material itself. The shape of the product crystals depends primarily on the character of the solvent and solute.

In crystallization the most important mechanism for particle size change is simply growth of the individual particles in the supersaturated solution.

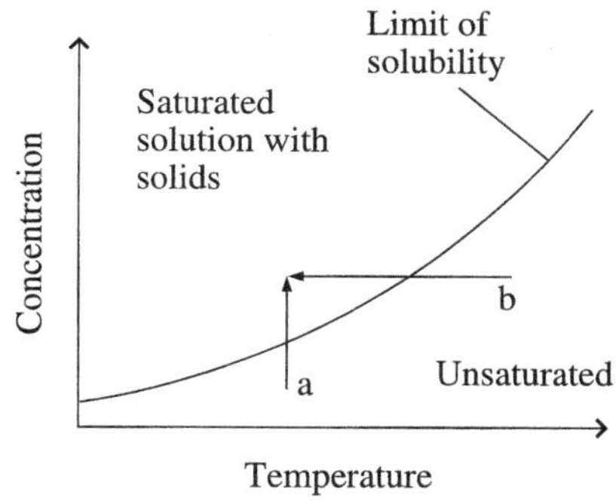

Figure 7.1: Illustration of crystallization. The same state of supersaturation may be brought about either by evaporation (a) or by cooling (b)

7.1.2 Agglomeration

In agglomeration processes particles are brought together into larger entities by agitation or pressing. Such processes are carried out in a broad range of processing industries for a variety of reasons.

We will concentrate on the process of granulation, where a number of particles are bound together into a granule, normally by adding a liquid binder while the particles are being agitated. Granulation of a powder consisting of fine particles may bring about many benefits, such as:

- improving the flow properties,
- reducing the porosity (void fraction),
- eliminating segregation, e.g. in a binary mixture where the primary particles have different physical properties and
- eliminating dustiness.

Figure 7.2 shows a sketch illustrating a granulation process.

Granulation is often carried out in either of two types of processing unit:

- a fluidized bed, to which the binder liquid is added from a spray nozzle, which may by suspended over the bed or be immersed in the bed; this is called "fluid bed granulation", or

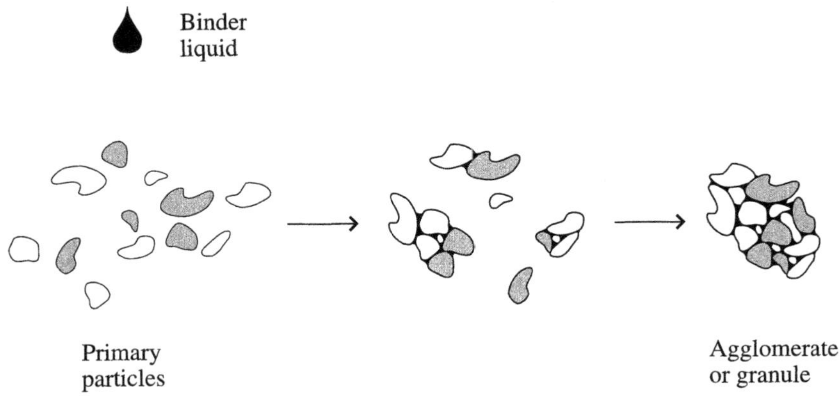

Figure 7.2: Sketch illustrating the agglomeration of particles with the help of a binder in a granulation process

- in a cylindrical mixer vessel equipped with an agitator, the whole thing looking a bit like a coffee grinder, to which the binder is added in spray form; this is called "high shear granulation".

As in crystallization, an important parameter determining product quality in granulation is the size distribution of the product granules, which in most granulation processes is determined primarily by the amount and type of binder added and a dynamic equilibrium between agglomeration and agglomerate break-up due to the agitation.

One example of PBM modeling in this chapter is a high-shear agglomeration process.

7.1.3 Comminution

Comminution, or grinding, which is particle size reduction by shredding or crushing, is a ubiquitous process, for instance in the energy, cement and mining industries. Coal may be crushed in order to control the particle size in a coal-fired furnace, or a large rock consisting of a non-homogeneous mixture of valuable ore and value-less gangue may be crushed to create smaller stones rich in ore or gangue, respectively, which can be separated from each other.

A central issue in the modeling of comminution processes is the size distribution of the particles produced by a breakage event. For instance, when a stone is crushed either by compression or by impact, it tends to break in the pattern indicated in Figure 7.3. Most of the stone is broken up in relatively coarse particles, while a cone of fine particles is produced in the vicinity of the point of impact. As a result, the particle size distribution of a powder produced in a mill acting by compression or impact tends to be bimodal. In fact, studying the size distribution of a given powder can yield quite a lot of information about how it was produced.

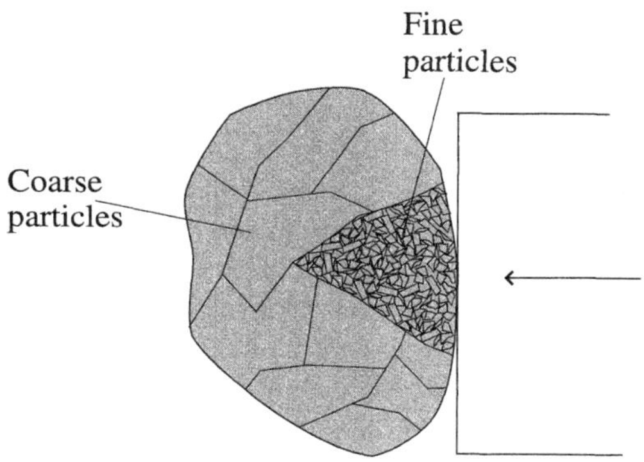

Figure 7.3: Sketch showing a typical break-up pattern as a result of crushing by impact or compression

Many types of mill are available for crushing and grinding operations, each with their advantages and disadvantages, and with a characteristic range of output particle sizes, see [101]. A comminution process will be the second example process that we will discuss in this chapter.

7.2 Basic Principles of Population Balance Modeling

We give a short introduction to the basic elements of the method here, and refer the reader to the two dedicated textbooks of A. D. Randolph and M. A. Larson [106] and D. Ramkrishna [105] for more information. We shall in this section be concentrating on the case where the size of the particle or agglomerate is the property of interest, but we will start with the more general case.

In the most general case a *vector of particle properties*, $\mathbf{x}$ is defined, consisting of

- an *internal property vector*, $\mathbf{x}_i$, with domain Ω_i bounded by the hypersurface Ω_i', where $\mathbf{x}_i$ may be the particle's size described by e.g. its diameter or its volume, the particle's surface area, its mass or its content of some specific chemical component, etc., and

- an *external property vector*, $\mathbf{x}_e$, with domain Ω_e bounded by the surface Ω_e'; the three components of $\mathbf{x}_e$ are simply the particle's spatial coordinates.

The internal properties, such as the particle size, are mostly continuous in nature, but in some processes may only take on only discrete values, such as in an agglomeration process of cells of a uniform size.

Further a *multidimensional number probability density function*, $f_n(\mathbf{x}) = f_n(\mathbf{x}_i, \mathbf{x}_e, t)$ is defined as the expectation, $E[n(\mathbf{x}_i, \mathbf{x}_e, t)]$, of the multidimensional number density function of the space of the internal and external properties. Instead of a number probability density function a volume or mass probability density funciton may be considered.

The motion of particles through multidimensional space may depend on external conditions. For a crystallization process such external conditions may be the temperature, the degree of supersaturation and the flowpattern of the solvent. These external conditions may themselves be functions of time and space.

Consistent with the physics of the processes under investigation, it is in PBM assumed that changes in the state of the particles may be either as a result of a smooth property transition, or as a result of a so-called "birth-death process", a term used in a particular sense in PBM. The two will be discussed separately.

7.2.1 Particle Nucleation and Growth

Smooth particle movement in internal property space, for example particle growth in a crystallization process and particle movement through physical space, can be represented by a "velocity", $\mathbf{v}$ with components in the space of internal properties, $\mathbf{v}_i = d\mathbf{x}_i/dt$ and in the space of external coordinates $\mathbf{v}_e = d\mathbf{x}_e/dt$. A component of $\mathbf{v}_i$ can, for instance, be the particle growth rate, $\partial(d_p)/\partial t$ while the vector $\mathbf{v}_e$ simply is the physical velocity of the particles in the reactor. $\mathbf{v}_i$ and $\mathbf{v}_e$ may depend on the particles' position in multidimensional space, $(\mathbf{x}_i, \mathbf{x}_e)$, for example on its size or on its position in the reactor, and on the external conditions, such as supersaturation, temperature or velocity of a carrier fluid.

Consider an arbitrary subdomain, $\Lambda(t)$, with boundary $\Lambda'(t)$ in our multidimensional space. Let the subdomain move with the particles, so that it contains a fixed number of particles:

$$\frac{d}{dt}\int_{\Lambda} f_n(\mathbf{x},t)dV = \frac{d}{dt}\int_{\Lambda_i} dV_i \int_{\Lambda_e} dV_e\ f_n(\mathbf{x}_i,\mathbf{x}_e,t) = 0. \tag{7.1}$$

We can treat our multidimensional space similarly to physical space [15, 17, 105, 106], and transform Equation (7.1) into a balance for a control volume with fixed coordinates, Λ_0, using an extension of Reynolds transport theorem:

$$\int_{\Lambda_{0,i}} dV_i \int_{\Lambda_{0,e}} dV_e \frac{\partial}{\partial t} f_n + \int_{\Lambda'_{0,i}} dS_i \int_{\Lambda'_{0,e}} dS_e \left(\mathbf{v}_i \cdot \mathbf{n} f_n + \mathbf{v}_e \cdot \mathbf{n} f_n\right) = 0, \tag{7.2}$$

where the last term on the left-hand-side represents particle flux over the hypersurface enclosing the control volume, and $\mathbf{n}$ is the unit normal vector pointing outward from the hypersurface Λ'. The argument of f_n has been dropped. Using an extension of Gauss' divergence theorem, this can be rewritten to:

$$\int_{\Lambda_{0,i}} dV_i \int_{\Lambda_{0,e}} dV_e \left[\frac{\partial}{\partial t} f_n + \nabla_i \cdot \mathbf{v}_i f_n + \nabla_e \cdot \mathbf{v}_e f_n\right] = 0. \tag{7.3}$$

Due to the arbitrariness of the control volume and therefore its size, we can drop the integrals, and write a differential population balance equation for smooth transitions:

$$\frac{\partial}{\partial t} f_n + \nabla_i \cdot \mathbf{v}_i f_n + \nabla_e \cdot \mathbf{v}_e f_n = 0. \tag{7.4}$$

To allow for spontaneous generation or removal of particles anywhere in the internal or external property space, for instance due to nucleation in a crystallization process or addition or removal of particles to or from the spatial domain, local rates of "birth", B, and "death", D, of particles per unit of multidimensional space are introduced:

$$\frac{\partial}{\partial t} f_n + \nabla_i \cdot \mathbf{v}_i f_n + \nabla_e \cdot \mathbf{v}_e f_n = B - D. \tag{7.5}$$

Addition or removal of particles may take place anywhere in the internal and external domains or through the boundaries of the domain.

Any boundary conditions for Equation (7.5) will depend on the specific application. Initial conditions may simply be some prescribed form of the multidimensional probability distribution function, f_n, over the domain.

Considerable simplifications are possible when applying Equation (7.5) to specific processes, including the examples we will consider later in this chapter. Considering such simple cases may aid the understanding. Let us, for example, consider a process where:

- The only internal property of interest is the particle size measured as its diameter, d_p.
- The process vessel is well mixed, so that the particles and all relevant parameters are homogeneously distributed in physical space and the size distribution of the outflowing particles is the same as that of the particles in the vessel.
- We have well-defined inflow and outflow streams.

In this case f_n is the one-dimensional size probability density function and $\mathbf{v}_i$ is simply the particle growth rate, which we can call k, with dimensions L/t. Due to the spatial homogeneity, gradients in the external coordinates—and therefore the third term on the LHS of Equation (7.5)—become zero. ∇_i becomes $\partial/\partial d_p$. The birth rate, B, is the sum of the inflow of particles to the vessel and the particle nucleation rate both per unit volume and per unit particle diameter, while the death rate, D, is the outflow of particles from the vessel per unit volume and per unit particle diameter. We call the volume of the reactor, which we identify with the spatial domain, Ω_e, V.

We can write the inflow and outflow rates in any given size band of width $d(d_p)$ as $Qf_{n,in}d(d_p)$ and $Qf_nd(d_p)$, respectively, where Q is the volumetric flowrate of the carrier liquid. If we assume that nucleated particles have a uniform size of $d_{p,0}$, then the nucleation rate per unit volume and per unit particle diameter can be written: $B_0\delta(d_p - d_{p,0})$ where δ, the Dirac delta function has the properties:

$$
\begin{aligned}
&\delta(x-a) = 0 && \text{for } x \neq a \\
&\delta(x-a) = \infty && \text{for } x = a \\
&\int_{a-\varepsilon}^{a+\varepsilon} f(x)\delta(x-a)dx = f(a) && \text{for } \varepsilon > 0,
\end{aligned}
$$

and B_0 is the nucleation rate in number of particles per volume per unit particle diameter.

We thus obtain from Equation (7.5) the following population balance equation for nucleation and growth in a well-mixed reactor with in- and outflow:

$$
\frac{\partial}{\partial t}f_n + \frac{\partial}{\partial d_p}kf_n = \frac{Q(f_{n,in} - f_n)}{V} + B_0\delta(d_p - d_{p,0}). \tag{7.6}
$$

A much-used technique to study the properties of Equation (7.6) is to study the *moments*, μ_j, of the particle size distribution, f_n. The j'th moment of an arbitrary function $f(x)$ with respect to its single argument, x is:

$$\mu_j = \int_0^\infty x^j f(x) dx.$$

For example, an ordinary differential equation for the j'th moment of f_n, μ_j, can be derived from Equation (7.6) for the case of size-independent growth rate, k, by taking the j'th moment of each of the terms with respect to the particle size, d_p:

$$\frac{\partial}{\partial t}\mu_j - jk\mu_{j-1} = \frac{Q}{V}(\mu_{j,in} - \mu_j) + B_0 d_{p,0}^j, \tag{7.7}$$

where we have integrated the second term on the left-hand-side of Equation (7.6) by parts:

$$\int_0^\infty d_p^j k \frac{\partial f_n}{\partial d_p} d(d_p) = [d_p^j k f_n]_0^\infty - \int_0^\infty j d_p^{j-1} k f_n d(d_p) = -jk\mu_{j-1},$$

and we have made use of the properties of the Dirac delta function given above.

The method of moments consists in solving (7.7) for some of the moments of f_n, rather than having to determine the whole of the distribution by solving (7.6). This is, as mentioned, a frequently used technique to which a voluminous research literature is dedicated. Moment equations for a variety of processes can be derived, see, for instance, D. Ramkrishna [105].

7.2.2 Birth-Death Processes of Agglomeration and Break-up

As indicated earlier in this chapter, it is in many contexts necessary to take into account processes wherein particles suddenly appear in or disappear from a given point in internal property space due to agglomeration or break-up. This is termed "birth-death" processes and modeled similarly to the above-mentioned processes of in- and outflow of particles and particle nucleation. In this section the word "particle" can refer to either a primary particle or an agglomerate.

Particle breakage

In general, the break-up of one particle gives rise to any number of particles with a distribution of "allowable" internal and external properties. An example of a non-allowable internal property for the particles resulting from a break-up is a size greater than that of the particle that broke.

Three "breakage functions" are normally defined:

- the fraction of particles with properties $(\mathbf{x}_{i,0}, \mathbf{x}_{e,0})$ breaking up per second, often called the "specific breakage rate function": $S(\mathbf{x}_{i,0}, \mathbf{x}_{e,0}, t)$,
- the average number of fragments produced in one break-up event at $(\mathbf{x}_{i,0}, \mathbf{x}_{e,0})$: $\nu_b(\mathbf{x}_{i,0}, \mathbf{x}_{e,0}, t)$ and
- The probability density function for the properties of a single particle produced by a break-up at $(\mathbf{x}_{i,0}, \mathbf{x}_{e,0})$: $P_b(\mathbf{x}_i, \mathbf{x}_e|\mathbf{x}_{i,0}, \mathbf{x}_{e,0}, t)$.

Each of these may also depend on the external conditions, such as the pressure or temperature or the composition of a solvent. The product of the two latter functions, $\nu_b(\mathbf{x}_{i,0}, \mathbf{x}_{e,0}, t)P_b(\mathbf{x}_i, \mathbf{x}_e|\mathbf{x}_{i,0}, \mathbf{x}_{e,0}, t) := b(\mathbf{x}_i, \mathbf{x}_e|\mathbf{x}_{i,0}, \mathbf{x}_{e,0}, t)$ is often called the "breakage distribution function", the integral of which with respect to $(\mathbf{x}_i, \mathbf{x}_e)$ over the entire multidimensional domain equals $\nu_b(\mathbf{x}_{i,0}, \mathbf{x}_{e,0}, t)$.

The total contribution to the birth rate at $(\mathbf{x}_i, \mathbf{x}_e)$ due to particle break-up becomes:

$$B_b = \int_{\Omega_i} dV_{i,0} \int_{\Omega_e} dV_{e,0}\, b(\mathbf{x}_i, \mathbf{x}_e|\mathbf{x}_{i,0}, \mathbf{x}_{e,0}, t)S(\mathbf{x}_{i,0}, \mathbf{x}_{e,0}, t)f_n(\mathbf{x}_{i,0}, \mathbf{x}_{e,0}, t), \tag{7.8}$$

and the total contribution to the particle death rate at $(\mathbf{x}_i, \mathbf{x}_e)$ due to particle break-up is:

$$D_b = S(\mathbf{x}_i, \mathbf{x}_e, t)f_n(\mathbf{x}_i, \mathbf{x}_e, t). \tag{7.9}$$

For the one-dimensional case, discussed in the previous section, of a well-mixed reactor wherein only the particle size is considered, the birth and death rates at d_b due to breakage become:

$$\int_0^\infty b(d_b|d_{b,0}, t)S(d_{b,0}, t)f_n(d_{b,0}, t)d(d_{b,0}) \tag{7.10}$$

and:

$$S(d_b, t)f_n(d_b, t), \tag{7.11}$$

respectively.

Particle Agglomeration

It is normally assumed that each particle agglomeration event takes place between only two particles or agglomerates.

We denote the number of distinct particle pairs in two differential volumes, the "pair-density function" dV_0 and dV_1 by $f_{n,p}(\mathbf{x}_{i,0}, \mathbf{x}_{e,0}, \mathbf{x}_{i,1}, \mathbf{x}_{e,1}, t)$ $dV_0 dV_1$, and the fraction of these particle pairs agglomerating per second as $\beta(\mathbf{x}_{i,0}, \mathbf{x}_{e,0}, \mathbf{x}_{i,1}, \mathbf{x}_{e,1}, t)$. β can also be regarded as the probability of a given particle pair agglomerating in a second (or the probability rate of agglomeration), which is often called the agglomeration rate constant or the agglomeration "kernel", and is sometimes broken up into the product of a collision rate for particle pairs and an "efficiency factor" for agglomeration, the latter is the probability that a given collision will lead to agglomeration. In what follows we will assume that β is independent of time and drop its time-argument.

We then obtain for the rate of births due to agglomeration, B_a, in a differential volume $dV = dV_i dV_e$ at $\mathbf{x} = (\mathbf{x}_i, \mathbf{x}_e)$:

$$\frac{1}{2}\int_{\Omega_i} dV_{i,0} \int_{\Omega_e} dV_{e,0}\ \beta(\mathbf{x}_{i,0}, \mathbf{x}_{e,0}, \mathbf{x}_{i,1}, \mathbf{x}_{e,1})$$
$$f_{n,p}(\mathbf{x}_{i,0}, \mathbf{x}_{e,0}, \mathbf{x}_{i,1}, \mathbf{x}_{e,1}, t) dV_{i,1} dV_{e,1} = B_a dV_i dV_e \quad (7.12)$$

where $(\mathbf{x}_{i,1}, \mathbf{x}_{e,1})$ are the specific particle properties required to form a particle at $(\mathbf{x}_i, \mathbf{x}_e)$ when agglomerating with one at $(\mathbf{x}_{i,0}, \mathbf{x}_{e,0})$. The factor of one half corrects for the redundancy arising from the integration counting all particle pairs twice.

Generally in this chapter we follow the tradition in the field of PBM of writing expressions separating the internal and external coordinates, but in this case we write Equation (7.12) in a simpler form in terms of the complete vector of property coordinates. Dividing through by $dV = dV_i dV_e$ this gives:

$$B_a = \frac{1}{2}\int_{\Omega} dV_0\ \beta(\mathbf{x}_0, \mathbf{x}_1) f_{n,p}(\mathbf{x}_0, \mathbf{x}_1, t)\frac{dV_1}{dV}. \quad (7.13)$$

The factor dV_1/dV converts a density with respect to $\mathbf{x}_1$ into one with respect to $\mathbf{x}$ [105]. This factor is usually equal to unity, but cannot be assumed to be unity in general.

The death rate at $(\mathbf{x}_i, \mathbf{x}_e)$ due to agglomeration, D_a, is:

$$D_a = \int_{\Omega_i} dV_{i,0} \int_{\Omega_e} dV_{e,0}\ \beta(\mathbf{x}_{i,0}, \mathbf{x}_{e,0}, \mathbf{x}_i, \mathbf{x}_e) f_{n,p}(\mathbf{x}_{i,0}, \mathbf{x}_{e,0}, \mathbf{x}_i, \mathbf{x}_e, t). \quad (7.14)$$

Normally the pair-density function may be calculated from:

$$f_{n,p}(\mathbf{x}_0, \mathbf{x}_1, t) = f_n(\mathbf{x}_0, t) f_n(\mathbf{x}_1, t).$$

Ramkrishna [105] discusses this issue further.

Again we can write the above expressions for a well-mixed reactor, where the internal property of interest is the particle size, d_p:

$$B_a = \frac{1}{2}\int_0^\infty d(d_{p,0})\ \beta(d_{p,0}, d_{p,1}) f_{n,p}(d_{p,0}, d_{p,1}, t). \tag{7.15}$$

and:

$$D_a = \int_0^\infty d(d_{p,0})\ \beta(d_{p,0}, d_p) f_{n,p}(d_{p,0}, d_p, t). \tag{7.16}$$

Further simplification of these two latter expressions is possible if the fact that the volume of an agglomerate equals the sum of the volumes of the contributing particles or agglomerates is utilized. We will consider this in the example of PBM discussed below.

7.3 An Example: PBM for High-Shear Granulation

In this section we will discuss an example of PBM modeling published in a recent paper by Biggs et al. [14]. The paper presents a model for the development with time of the agglomerate size in a high-shear granulation process.

7.3.1 The Model

As Figure 7.2 indicates, an agglomerate or granule[1] will consist of three phases: solid, liquid and gas. The rates of the processes taking place during the granulation—such as agglomeration, breakage and agglomerate compaction—will depend on the size and composition of the agglomerates at any given time. For instance, the higher the liquid content of an agglomerate, the more easily it will agglomerate with another or incorporate additional primary particles.

In principle, therefore, a population balance model with a multidimensional internal property vector is necessary for granulation. Furthermore, the processes may depend on the position of the agglomerate in the granulator vessel, so that also a three-dimensional external property vector has to be taken into account.

We assume that the gas rather quickly escapes a formed agglomerate, so that the only two internal properties we need to take into account are

[1] We use these two words as alternatives here, tending to use the work "granule" when wishing to emphasize the aspect of product from the process

the size and the liquid content. Furthermore, we assume the granulator to be well-mixed and homogeneous, so that we do not need to account for an external property vector, i.e. the spatial coordinates of the particle in the vessel.

We also assume the process to be a *pure agglomeration process*, such that we do not need to take particle/agglomerate breakage or particle growth into account.

We thus obtain from Equations (7.12) and (7.14):

$$\frac{\partial f_n(v, v_L, t)}{\partial t} = B_a - D_a = B - D =$$
$$\frac{1}{2}\int_{\Omega_i} dV_{i,0}\ \beta(\mathbf{x}_{i,0}, \mathbf{x}_{i,1}) f_{n,p}(\mathbf{x}_{i,0}, \mathbf{x}_{i,1})$$
$$- \int_{\Omega_i} dV_{i,0}\ \beta(\mathbf{x}_{i,0}, \mathbf{x}_i) f_{n,p}(\mathbf{x}_{i,0}, \mathbf{x}_i, t), \quad (7.17)$$

where we have also assumed that number density with respect to $\mathbf{x}_0$ is equal to density with respect to $\mathbf{x}$.

We take the granule volume, v, to be the measure of size and the volume of liquid in a granule v_L to be a measure of the liquid content. The two-dimensional number density function is $f_n(v, v_L, t)$.

With these measures the population balance can be further simplified, since we know that a particle or agglomerate of internal properties $\mathbf{x}_0 = (\varepsilon, \gamma)$ must agglomerate with one of internal properties $\mathbf{x}_1 = (v - \varepsilon, v_L - \gamma)$ to produce one with internal properties $\mathbf{x} = (v, v_L)$. Moreover, we assume that the agglomeration kernel, β, only depends on the size of the particle/agglomerate and/or some property that is determined by only the size.

We thus obtain:

$$\frac{\partial f_n(v, v_L, t)}{\partial t} =$$
$$\frac{1}{2}\int_0^\infty d\varepsilon \int_0^\infty d\gamma\ \beta(v - \varepsilon, v_L - \gamma, \varepsilon, \gamma) f_n(v - \varepsilon, v_L - \gamma, t) f_n(\varepsilon, \gamma, t)$$
$$- f_n(v, v_L, t) \int_0^\infty d\varepsilon \int_0^\infty d\gamma\ \beta(v, v_L, \varepsilon, \gamma) f_n(\varepsilon, \gamma, t). \quad (7.18)$$

We note that the integrand of the first term on the right-hand-side is zero if ε exceeds v or if γ exceeds v_L or ε, and that the integrand in the second term on the right-hand-side is zero if γ exceeds ε.

For a rigorous study of agglomeration during the granulation process, Equation (7.18) needs to be solved for the two-dimensional density function f_n. However, the problem may be simplified to involve two density functions each in one dimension. These are the number density function for granule volume:

$$n(v,t) = \int_0^\infty f_n(v, v_L, t) dv_L$$

and the mass density function for liquid content in a granule:

$$M(v,t) = \rho_L \int_0^\infty v_L f_n(v, v_L, t) dv_L.$$

If we further assume that v_L is a function of v, so that all granules of a given size have the same liquid content, then β is no longer an explicit function of v_L, and we can write:

$$\frac{\partial n(v,t)}{\partial t} = \frac{1}{2}\int_0^\infty d\varepsilon\ \beta(v-\varepsilon,\varepsilon)n(v-\varepsilon,t)n(\varepsilon,t) - n(v,t)\int_0^\infty d\varepsilon\ \beta(v,\varepsilon)n(\varepsilon,t). \quad (7.19)$$

This implies an equation for the mass distribution of liquid:

$$\frac{\partial M(v,t)}{\partial t} = \frac{1}{2}\int_0^\infty d\varepsilon\ \beta(v-\varepsilon,\varepsilon)M(v-\varepsilon,t)n(\varepsilon,t) - M(v,t)\int_0^\infty d\varepsilon\ \beta(v,\varepsilon)n(\varepsilon,t). \quad (7.20)$$

Here we recognize that for physical reasons the integrands are zero if v_L exceeds v.

7.3.2 Solution Method

A numerical solution method is used, first described by Hounslow et al. [65].

The size domain is discretized at points $(v_i)_{i=1\ldots\infty}$ forming a geometric series such that

$$\frac{v_i}{v_{i-1}} = 2.$$

The interval i is $[v_i, v_{i+1})$, and contains $N_i = n_i(v_{i+1} - v_i)$ particles, which are assumed to be homogeneously distributed, so that the particle number density, n_i, is uniform over the interval i.

Since the analysis below is based only on the nature of this geometric series and not on the absolute values of the v_i, it can, for the derivation of the discretized population balance equation, be assumed that the $v_i = 2^i$ without loss of generality. With these values, the width of the i'th interval is 2^i.

Consider first the rate of births in interval i. In this pure agglomeration process a birth in interval i occurs if and only if a particle in interval $i-1$ participates in an agglomeration event.

If the other participating particle is also in interval $i-1$, then the event will always result in a birth in i, such that the frequency due to this is:

$$B_{i,1} = \frac{1}{2}\beta_{i-1,i-1}N_{i-1}N_{i-1}$$

where the factor one half avoids counting particle pairs twice.

If the other participating particle, however, has a volume ξ, which is in interval $j < i-1$, the agglomeration only results in a birth in i if the volume ζ of the particle in interval $i-1$ with which it agglomerates is such that: $v_i - \xi < \zeta < v_i$, that is: $2^i - \xi < \zeta < 2^i$. There are ξn_{i-1} such particles. The total number of agglomeration events causing births in i, and involving particles in a band dv around ξ is then:

$$\beta_{i-1,j}\xi n_{i-1}n_j dv = \beta_{i-1,j}\frac{\xi N_{i-1}}{2^{i-1}}\frac{N_j}{2^j}dv.$$

The total number of agglomeration events between particles in $j < i-1$ and $i-1$ causing births in i is then obtained by integrating over the j'th interval:

$$\beta_{i-1,j}\int_{2^j}^{2^{j+1}} dv\ v\ 2^{1-i-j}N_{i-1}N_j = 3 \times 2^{j-i}\beta_{i-1,j}N_{i-1}N_j.$$

The total rate of births in i due to agglomeration between particles in $i-1$ and particles in $j = 1 \ldots i-2$ is then obtained by summation over j:

$$B_{i,2} = \sum_{j=1}^{i-2} 3 \times 2^{j-i}\beta_{i-1,j}N_{i-1}N_j.$$

Turning now to the rate of deaths in interval i due to agglomeration, a death in interval i will result if a particle in i agglomerates with one in interval $j \geq i$. The total rate of deaths due to this is therefore:

$$D_{i,1} = N_i\sum_{j=i}^{\infty}\beta_{i,j}N_j.$$

If, however, the particle of volume ζ in i agglomerates with a particle of volume ξ in $j < i$, a death will only occur if the volume of the resulting agglomerate exceeds $v_{i+1} = 2^{i+1}$, i.e. if $2^{i+1} - \xi < \zeta < 2^{i+1}$. The total number agglomeration events causing deaths in i, and involving particles in a band dv around ξ is then:

$$\beta_{i,j}\xi n_i n_j dv = \beta_{i,j}\frac{\xi N_i}{2^i}\frac{N_j}{2^j}dv.$$

The total number of agglomeration events between particles in $j < i$ and i causing deaths in i is then obtained by integrating over the j'th interval:

$$\beta_{i,j}\int_{2^j}^{2^{j+1}} dv\ v\ 2^{-i-j}N_iN_j = 3 \times 2^{j-i-1}\beta_{i,j}N_iN_j.$$

The total rate of deaths in i due to agglomeration between particles in i and particles in $j = 1 \ldots i-1$ is then obtained by summation over j:

$$D_{i,2} = \sum_{j=1}^{i-1} 3 \times 2^{j-i-1}\beta_{i,j}N_iN_j.$$

The overall rate of change of particles in interval i may then be obtained by subtracting the two contributions to the death rate from the two contributions to the birth rate. However, it turns out that while such a scheme conserves the number of particles in the system it does not conserve the total particle volume. $B_{i,1}$ and $D_{i,1}$ are therefore multiplied by a "volume correction factor" of 2/3 to cause both number and volume to be conserved (see Hounslow et al [65]). Thus:

$$B_i - D_i = \frac{dN_i}{dt} = \frac{2}{3}B_{i,1} + B_{i,2} - \frac{2}{3}D_{i,1} - D_{i,2}$$

giving the final discretized population balance equation for the agglomeration process:

$$\frac{dN_i}{dt} = \frac{1}{2}\beta_{i-1,i-1}N_{i-1}^2 + N_{i-1}\sum_{j=1}^{i-2}\left(2^{j-i+1}\beta_{i-1,j}N_j\right) - N_i\sum_{j=i}^{\infty}\beta_{i,j}N_j - N_i\sum_{j=1}^{i-1}\left(2^{j-i}\beta_{i,j}N_j\right) \quad (7.21)$$

where the upper limit of the sum in the third term on the right-hand-side in practice is replaced by $\mathcal{N}$, the maximal size interval considered.

In the article of Biggs et al. [14], a discretized version of Equation (7.20) is also given, based on a derivation presented in Hounslow et al. [66]. This second equation is solved together with Equation (7.21) to determine the liquid content of agglomerates during the process, a parameter which, as mentioned, may be important for the probability of agglomeration upon collision, and therefore for the agglomeration kernel, β.

We will not do this here, where our primary objective is to give an example of PBM. We simply approximate the data for liquid content as a function of granule size given in the paper of Biggs et al. (their Figure 1) by a simple functional form, as shown in the following section.

7.3.3 Quantification of Parameters and Comparison with Experiment

Biggs et al. compare model predictions with experimental results for high-shear granulation of two powders: calcium carbonate and lactose. We will compare the predictions of our simplified models (modeling the particle size, not the liquid content) with the results for calcium carbonate.

Quantification of the Parameters

Various forms for the agglomeration kernel, β, are proposed in the literature. It is assumed here that the kernel depends on the liquid content of the colliding particles only, and therefore indirectly only on the size. We can thus write:

$$\beta_{i,j} = \beta_0 \psi_{i,j}(\Phi_{l,i}, \Phi_{l,j}),$$

where β_0 accounts for the collision frequency and overall agglomeration probability, and $\psi_{i,j}(\Phi_{l,i}, \Phi_{l,j})$ for the dependency of the agglomeration probability on the volume fraction of liquid in the agglomerating particles/agglomerates, $\Phi_{l,i}$ and $\Phi_{l,j}$.

Biggs et al. assume that a critical liquid content, Φ_c is necessary for agglomeration to take place at all, ψ can therefore be approximated by a logistic function:

$$\psi_{i,j} = \frac{(\max[\Phi_{l,i}, \Phi_{l,j}]/\Phi_c)^{30}}{1 + (\max[\Phi_{l,i}, \Phi_{l,j}]/\Phi_c)^{30}}.$$

We have decreased the sharpness of the logistic function slightly compared to that used in Biggs et al., using a power of 30 rather than the 48 to avoid the minimum in the resulting curves, shown below, to become too sharp. It is not physically unrealistic that the transition in ψ from zero to unity should not be too sudden.

As mentioned, rather than solving a separate equation for the agglomerate liquid content, we roughly approximate the data for liquid volume fraction as a function of agglomerate size of Biggs et al. at 240 s with the simple expression:

$$\Phi_l = \begin{cases} 0.15 + (0.40 - 0.15)\dfrac{d_p}{1.0} & \text{for } d_p \leq 1.0 \\ 0.40 + (0.35 - 0.40)\dfrac{d_p - 1.0}{3.0} & \text{for } d_p > 1.0 \end{cases}$$

and assume this to be constant throughout the granulation process, an assumption which is reasonable from a physical point of view, and to an extent is borne out by the results of Biggs et al.

Comparison with Experiment

The experimental procedure is described in Scott et al. [113]. The feed powder to the high-shear granulator was calcium carbonate "Durcal 40", and the liquid binder was polyethylene glycol. The mass of particles in each size class as a function of time was determined by sieve analysis of samples obtained from the running process by periodically stopping the mixer and obtaining samples of 60–70 g, which were subsequently cooled from the 60°C operating temperature through repeated passes through a sample divider. The fraction of binder liquid in the granules was determined by thermogravimetric analysis.

Biggs et al. solved the discretized population balances using the Runge-Kutta option in Mathematica. True to the spirit of this book, we simply discretized time, and calculated the population of each size class in time interval m using the simple explicit scheme:

$$\begin{aligned} N_i(m) = \Delta t \Bigg[& \frac{1}{2}\beta_{i-1,i-1}N_{i-1}^2(m-1) \\ & + N_{i-1}(m-1)\sum_{j=1}^{i-2}\left(2^{j-i+1}\beta_{i-1,j}N_j(m-1)\right) \\ & - N_i(m-1)\sum_{j=i}^{\mathcal{N}}\beta_{i,j}N_j(m-1) - N_i(m-1)\sum_{j=1}^{i-1}\left(2^{j-i}\beta_{i,j}N_j(m-1)\right)\Bigg]. \end{aligned} \quad (7.22)$$

A time-step of 1 s was a sufficiently fine discretization, giving results hardly distinguishable from those obtained using a time step of 0.1 s.

The computations were, as in the paper of Biggs et al., initiated with the experimental results for the N_i at a time of 240 s.

Figure 7.4 shows the experimental results together with the model predictions. We did not perform a formal optimization of the model parameters like Biggs et al. did but we did manually determine near-optimal parameters. These near-optimal parameter values were: $\Phi_c = 0.25$ mm and $\beta_0 = 5.0 \times 10^{-10}$, and are thus very close to the values obtained in the more complete model of Biggs et al. of $\Phi_c = 0.215$ mm and $\beta_0 = 3.35 \times 10^{-10}$.

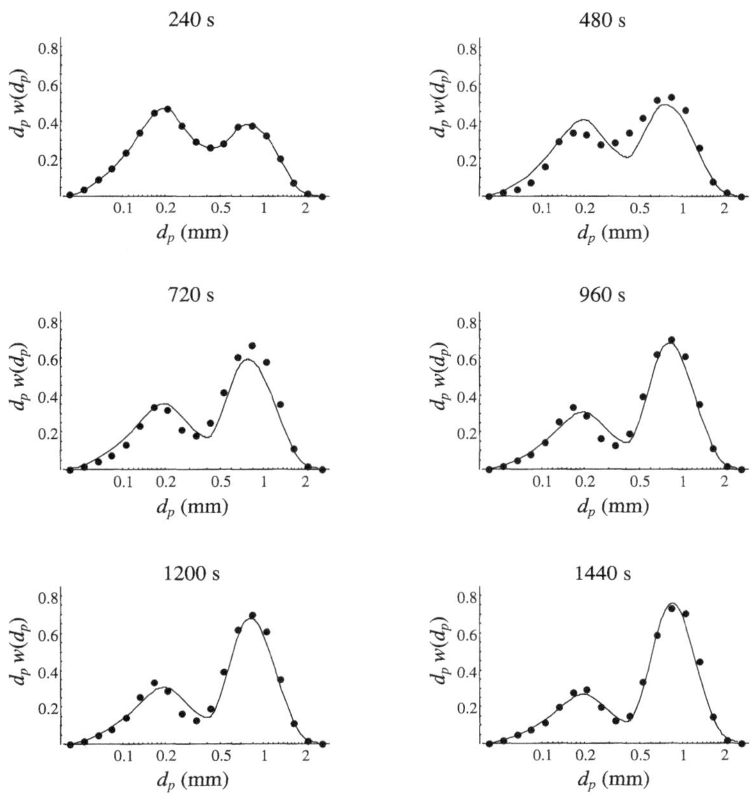

Figure 7.4: Agglomerate size distribution at successive times in the high-shear granulation of calcium-carbonate powder. Points are the experimental results and lines are the model predictions

We see from the figure that the simplified analysis here, assuming that the granule liquid content is a fixed function of the granule size throughout the process, also yields an acceptable fit between the model predictions and the experimental results.

7.3.4 Relation Between PBM and Stochastic Markovian Modeling of Agglomeration

We mentioned earlier in this chapter that PBM modeling, although based on a stochastic formulation, is mostly applied in an essentially deterministic way with reference to the law of large numbers. The PBM equations given above are for the expectation of the population distribution, just as most of the other models in this book. As we have pointed out several times, however, this does not detract from the advantages of using stochastic modeling principles during the model formulation and the model solution stages.

An additional advantage of stochastic modeling techniques is the possibility of obtaining information about the statistical spread in the population distribution. This is especially useful for small populations, such as in the small-scale, highly intense processes that are increasingly being developed, and in processes involving small mixing zones or significant inhomogeneities on a scale comparative to that of the particles. Moreover, in processes, for example in the nourishment or medical industries, where the object is to eradicate some potentially health-threatening agent or evenly distribute some medically active agent in a mixture the statistical spread around the mean population may be a crucial parameter.

D. Ramkrishna [105] has developed the statistical foundation for population balances in chapter 7 of his dedicated textbook.

As in other contexts in this present book, the question arises whether PBM model equations of the type described above can be cast in the form of a discretized Markovian process model by formulation of a transfer probability matrix which can be solved to yield population probability vectors at successive time intervals.

This should in principle be possible. Equation (7.22) shows that the process is Markovian, since the population probability distribution can be calculated from only the population probability in the previous time step. A renumbering of the cells in discretized multi-dimensional space, consisting of the internal and external property spaces, similar to that performed in Chapter 6.1.1 will make it possible to operate with a probability vector for a stochastic variable, which could reflect the population distribution probability of some conserved quantity such as, for the agglomeration process described here, the mass of particles/agglomerates.

The transfer probability matrix, however, would be time-dependent, depending, as it does, on the current probability vector, and its formulation would be a computational challenge, similar to the challenge of solving the discretized population balance equations above.

In the following section we will describe work wherein traditional PBM modeling has been linked directly to stochastic modeling by Markov chains for a comminution process.

7.4 Analysis of a Grinding Process

In this section we will give an account of the work of Berthiaux [8] aimed at analysing a grinding (or comminution) process by Markov chains, and at linking this type of analysis to traditional PBM analysis.

7.4.1 Traditional modeling by PBM

We have already discussed the physics of particle breakup or comminution in Section 7.1.3 and the general structure of its PBM modeling in Section 7.2.2.

The PBM analysis of Berthiaux is based on a discrete PBM analysis in which he discretizes property space. The only property he analyses is the particle size expressed as the diameter. He measures the amount of particles in each of the n size classes as the mass fraction, m_i, of particles.

He therefore defines two breakage matrices as follows (see also Berthiaux and Dodds [9]):

- The "breakage function", $\mathbf{B}$, an $n \times n$ matrix with elements b_{ij} that represent the mass fraction of those particles of size class j that break during a short (infinitesimal) interval of grinding that enter size class i. In other words: b_{ij} is the probability of material moving to class i conditional upon it exiting class j through grinding. In yet other words: b_{ij} is the distribution over the size classes of that material that through grinding exits class j. This is a discrete equivalent of the density function we called the "breakage distribution function" in Section 7.2.2.

- The "selection function", a column vector of length n with elements S_i, which represent the mass fraction of particles in size class i that *per second* moves to another size class during a short (infinitesimal) interval of grinding. This is a discrete version of the 'probability density function for the properties of a single particle produced by a break-up in Section 7.2.2.

With these definitions, the change with time of the mass fraction of material in size class i, m_i becomes:

$$\frac{dm_i}{dt} = -S_i m_i + \sum_{j=1}^{i} b_{ij} S_j m_j. \tag{7.23}$$

Equation (7.23) can be written in matrix form:

$$\frac{d\mathbf{M}}{dt} = (\mathbf{B} - \mathbf{I})\mathbf{S}\mathbf{M} = \mathbf{A}\mathbf{M} \tag{7.24}$$

where $\mathbf{M}$ is the column vector of length n with elements m_i, $\mathbf{I}$ is the identity matrix and $\mathbf{S}$ is a diagonal $n \times n$ matrix containing the elements S_i. We have defined the matrix $\mathbf{A}$ with the last equality.

Equation (7.24) is a homogeneous system of first order differential equations and can be solved by the standard technique involving matrix diagonalization (see Chapter 2 and Kreyszig [79]):

$$\mathbf{M}(t) = \mathbf{P}\mathbf{E}_D(t)\mathbf{P}^{-1}\mathbf{M}(0) = \mathbf{V}(t)\mathbf{M}(0) \tag{7.25}$$

where $\mathbf{P}$ is the matrix of $\mathbf{A}$'s eigenvectors, and the factor $\mathbf{P}^{-1}\mathbf{M}(0)$ is the integration constants. We have defined $\mathbf{V}(t)$ with the last equality.

The matrix $\mathbf{E}_D$ is the diagonal matrix with elements $\exp(\lambda_i t)$, where λ_i are the eigenvalues of $\mathbf{A}$. $\mathbf{B}$ is lower triangular and its diagonal elements are zero, since material only moves to smaller size classes due to grinding. Since, moreover, $\mathbf{S}$ is diagonal, Equation (7.24) shows $\mathbf{A}$ to be lower triangular with the diagonal elements, equal to $-S_i$, its eigenvalues. $\mathbf{E}_D$ thus has the elements $\exp(-S_i t)$.

7.4.2 Modeling by Markov chains

Berthiaux defines the discrete state space of a Markov chain as the set of possible size classes (n in total, where the index n refers to the smallest size class) for the particles. The Markov chain is discrete also in time, and is defined by its transition matrix $\mathbf{T}_m$ describing the change in the state of material due to grinding during one time interval of length Δt.

The elements of the transition matrix $\mathbf{T_m}$, t_{ij} are thus defined as the probability of material being in size class j in time-step $p+1$, conditional on the material having been in class i in time-step p. $\mathbf{T}_m$ is upper triangular, since the grinding can only lead to transfer to a lower size class.

$$\mathbf{T}_m = \begin{pmatrix} t_{11} & t_{12} & \cdots & \cdots & t_{1n} \\ 0 & t_{22} & & & \vdots \\ \vdots & \ddots & \ddots & & \vdots \\ \vdots & 0 & \ddots & \ddots & \vdots \\ 0 & \cdots & \cdots & 0 & t_{nn} \end{pmatrix}.$$

The probability vector, $\boldsymbol{\pi}_p$, with elements summing to unity, where the index p denotes the time interval, describes the probability distribution over the states of marked material in the grinding process.

Similarly to, e.g. our models for fluidized beds, the evolution of the probability vector can be evaluated by matrix multiplication:

$$\boldsymbol{\pi}_1 = \boldsymbol{\pi}_0 \mathbf{T}_m \tag{7.26}$$

and:

$$\boldsymbol{\pi}_p = \boldsymbol{\pi}_0 (\mathbf{T}_m)^p. \tag{7.27}$$

As described in Chapter 2 finding $(\mathbf{T}_m)^p$ can be done by diagonalization. $\mathbf{T}_m$ can be written:

$$\mathbf{T}_m = \mathbf{R}\mathbf{D}\mathbf{R}^{-1} \tag{7.28}$$

where $\mathbf{R}$ is the matrix of the eigenvectors of $\mathbf{T}_m$ and $\mathbf{D}$ is the diagonal matrix containing the eigenvalues of $\mathbf{T}_m$, which, since $\mathbf{T}_m$ is triangular, are the elements on its diagonal. We obtain:

$$\mathbf{D} = \begin{pmatrix} t_{11} & 0 & \cdots & \cdots & 0 \\ 0 & t_{22} & & & \vdots \\ \vdots & \ddots & \ddots & & \vdots \\ \vdots & 0 & \ddots & \ddots & \vdots \\ 0 & \cdots & \cdots & 0 & t_{nn} \end{pmatrix}.$$

and:

$$\mathbf{R} = \begin{pmatrix} 1 & r_{12} & \cdots & \cdots & r_{1n} \\ 0 & 1 & & & \vdots \\ \vdots & \ddots & 1 & \ddots & \vdots \\ \vdots & & \ddots & 1 & r_{n-1,n} \\ 0 & \cdots & \cdots & 0 & 1 \end{pmatrix}.$$

where the elements of $\mathbf{R}$ are found from:

$$r_{ij} = \sum_{k=i+1}^{j} \left(\frac{t_{ik}}{t_{jj} - t_{ii}} \right) r_{kj}.$$

and thus

$$\boldsymbol{\pi}_p = \boldsymbol{\pi}_0 \mathbf{R} \mathbf{D}^p \mathbf{R}^{-1} \tag{7.29}$$

-see also Equation (2.6).

7.4.3 Comparison of the Two Models

Both of the vectors $\boldsymbol{\pi}_p$ and $\mathbf{M}(t)$ contain the mass fraction of a pulse of marked material in the process. Comparing Equations (7.25) (7.27) shows that if the two vectors are equal, i.e. the two models give the same result, then:

$$(\mathbf{T}_m)^p = \mathbf{V}(p\Delta t)^{\mathrm{T}} \tag{7.30}$$

where $\cdot^{\mathrm{T}}$ denotes the transpose.

Berthiaux goes further to compare the approaches in more detail by comparing the breakage functions of PBM to the transfer probability matrix of the Markov-chain model.

To do this, he assumes all the marked material to be concentrated in size class j and time t, and denotes its mass fraction by $m_j^*(t)$. He denotes the mass fraction of marked material in class i after one time step of length Δt as $m_i^*(t+\Delta t)$. The elements of the transfer probability matrix can then be written:

$$t_{ji} = \frac{m_i^*(t + \Delta t)}{m_j^*(t)}. \tag{7.31}$$

If the time-step, Δt, is small enough we can estimate the breakage functions from first-order approximations (refer to the definitions of b_{ij} and S_i in Section 7.4.1):

$$b_{ij} = \frac{m_i^*(t + \Delta t)}{m_j^*(t) - m_j^*(t + \Delta t)} \tag{7.32}$$

and:

$$S_j = \frac{m_j^*(t) - m_j^*(t + \Delta t)}{m_j^*(t)\Delta t} = \frac{1}{\Delta t} \left(1 - \frac{m_j^*(t + \Delta t)}{m_j^*(t)} \right). \tag{7.33}$$

Comparing Equations (7.31), (7.32) and (7.33) gives Berthiaux the following relationships between the breakage functions and the transfer probability matrix:

$$b_{ij} = \frac{t_{ji}}{1 - t_{jj}} \tag{7.34}$$

and

$$S_j = \frac{1}{\Delta t}(1 - t_{jj}). \tag{7.35}$$

Solving Equations (7.34) and (7.35) for t_{ji} and t_{jj} gives:

$$t_{ji} = b_{ij} S_j \Delta t \tag{7.36}$$

and

$$t_{jj} = 1 - S_j \Delta t, \tag{7.37}$$

which can be used to fill in the transfer probability matrix in terms of the particle breakage functions.

In fact, the matrix $\mathbf{T}_m$ can be written as a function of the $\mathbf{B}$ and $\mathbf{S}$ matrices:

$$\mathbf{T}_m = \mathbf{I} - \Delta t\mathbf{S} + \Delta t\mathbf{S}\mathbf{B}^{\mathrm{T}} = \mathbf{I} + \Delta t\mathbf{A}^{\mathrm{T}} \tag{7.38}$$

Diagonalizing $\mathbf{A}$ (Chapter 2), and taking into account, as stated above, that the eigenvalues of $\mathbf{A}$ are $-S_i$ gives:

$$\mathbf{T}_m = \mathbf{I} - \Delta t(\mathbf{P}^{-1})^{\mathrm{T}}\mathbf{S}\mathbf{P}^{\mathrm{T}}. \tag{7.39}$$

Equation (7.37) shows that $t_{jj} = 1 - \Delta t S_j$, so that $\mathbf{D} = \mathbf{I} - \Delta t\mathbf{S}$, so that Equation (7.28) may be written:

$$\mathbf{T}_m = \mathbf{I} - \Delta t\mathbf{R}\mathbf{S}\mathbf{R}^{-1} \tag{7.40}$$

Comparing Equations (7.39) and (7.40), we see that:

$$R = (P^{-1})^{\mathrm{T}}$$

Obviously the above developments, and the entire concept of modeling the grinding or comminution process with a stationary Markov chain model discretized in time, depends on:

a) The breakage functions being independent of time.

b) The time step, Δt, being small enough for the first-order estimates of b_{ij} and S_j to be good enough.

Ad a): This may not be the case in practice, which is discussed extensively in the research literature on comminution.

Ad b): It was stated in Equation (7.30) in the beginning of this section that if the two models gave the same result, $(\mathbf{T}_m)^p$ would be equal to $\mathbf{V}(p\Delta t)^{\mathrm{T}}$. From another point of view, if the two are sufficiently close, the model results will be sufficiently close. Comparing the diagonal elements of the two matrices written in terms of the S_i, this means that $1 - \Delta t S_i$ has to be sufficiently close to $\exp -\Delta t S_i$.

Berthiaux provides a discussion of the advantages of the Markov chain model in comparison with the standard PBM model. One major advantage is that the direct experimental determination of the transfer probability matrix[2] is much easier than having to determine the breakage matrices experimentally. Berthiaux uses an example to show that the time-discretization for the Markov chain model can be rather rough for an acceptable agreement with the time-continuous PBM model, something that we have also found a number of times in this book, and which allows for very economical evaluation of the stochastic model.

[2] In the paper we have discussed in this section, Berthiaux determined the transfer probability matrix from the breakage matrices to show the link between PBM and Markov chain modeling, but normally one would determine the elements of the transfer probability matrix directly from experiment.

Chapter 8

Multiphase Systems

In Chapters 3 and 5 Markov chain models for bubbling fluidized beds have been introduced. These models consisted of a Markov birth-death chain with an additional jump parameter to the top (or underneath a baffle) λ_i governing the extra transport upward of particles caught in the wake of a rising gas bubble. Introducing such a jump to the top which amounts to admitting infinite velocity for those rising particles was certainly a model assumption of which feasibility and impact had still to be examined, although very good results had already been achieved with this approach.

Recent experiments by Dechsiri et al. [33] using a PET-camera show a quite fast rise of particles to the top but nonetheless one with finite velocity (see Figure 8.7 later in this chapter) which explains the quality of results of the model with jumps. However, this feature should be included in models dealing with these kind of processes . Unfortunately a simple extension of the existing models is not possible since the necessary information if a particle is drifting and diffusing downward with some velocity or moving upward in the wake of a rising gas bubble with some other velocity is not preserved in the actual state, i.e. the particles location in the reactor. Thus the process loses its Markovian character.

In this chapter we show how to construct a new Markov process capturing this additional feature and still giving easy access to the point of interest at hand, i.e. the particle's location, by introducing a new so-called multiphase process. This is a general concept that can be used for a wide variety of applications, e.g. to model different chemical phases (see Section 6.2) or different sizes in Population Balance Modeling (PBM, see Chapter 7). In style with this book and to further the easy understanding of such a multiphase process or multiphase system we develop its theory on the basis of an example namely the stochastic modeling of fluidized beds as in Chapter 5. An extension of the presented multiphase system is given such that it also incorporates the feature of exchange of wake material between bubble and bulk flow and and the phenomenon of gulf streaming. Afterwards the model is validated by comparing its predictions with empirical data. In the sparse literature about multiphase systems the article of Too et al. [119] stands out. A brief review of their work and a discussion especially concerning similarities and differences to our approach in this chapter is presented. We also include a consideration of using the necessary conditions for Danckwerts' law to hold as introduced in Chapter 4 in such a case of a multiphase system. Finally we briefly summarize the modeling of multiphase systems with a more abstract and general approach. The formulation of a stochastic model framework for multiphase systems presented in this chapter increases significantly the range of systems and processes that can be modeled, while maintaining the Markovian character of the model.

8.1 Multiphase System for a Bubbling Fluidized Bed

We consider a bubbling fluidized bed and its stochastic model using a Markov chain as introduced in Chapter 5 and shown in Figure 8.1. To give a short recap of the Markov model we present the fundamental facts. The

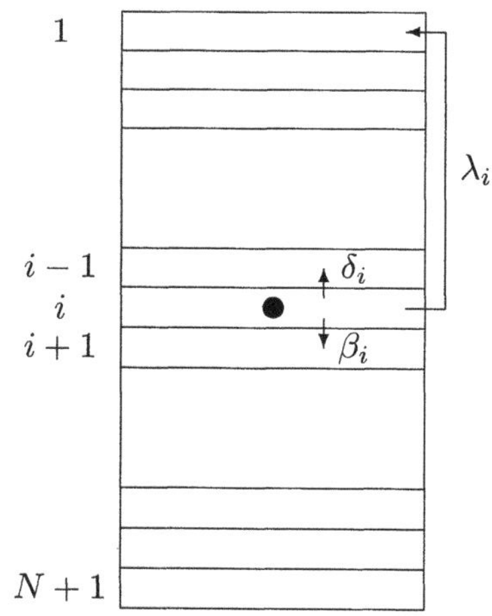

Figure 8.1: Discretized fluidized bed with arrows indicating all possible transition paths for the particle located in cell i.

bed is discretized in layers numbered from 1 to N and an exit (absorbing) state with the label $N+1$. The transition matrix $\mathbf{P}$ consists of probabilities for $2 \leq i \leq N$:

$$\begin{aligned} p_{i,i} &= \alpha_i(1-\lambda_i) = (1-\beta_i-\delta_i)(1-\lambda_i), \\ p_{i,i+1} &= \beta_i(1-\lambda_i), \\ p_{i,i-1} &= \delta_i(1-\lambda_i), \\ p_{i,1} &= \lambda_i \end{aligned}$$

and

$$\begin{aligned} p_{1,1} &= (1-\beta_1)(1-\lambda_1), \\ p_{1,2} &= \beta_1(1-\lambda_1), \\ p_{N+1,N+1} &= 1, \end{aligned}$$

at the boundaries. The object under examination is the particle's location at each timestep $n \geq 0$ denoted by the random variable $X_n \in \{1, 2, \ldots, N+1\}$.

8.1.1 The Multiphase System

To introduce a second phase with upward movement to replace the jumps and retain the Markovian character of the model we have to keep all necessary information to formulate a Markov chain. Therefore we have to enlarge the state space. At each timestep the probability of the upcoming transition depends on the location of the particle and its present phase. As phases we distinguish in this example process the ordinary downward drift with diffusion and the fast rise upward in the wake of a gas bubble. This leads to the following state space

$$S = \{1, 2, \ldots, N\} \times \{0, 1\} \cup \{(N+1, 0)\}.$$

With a slight abuse of notation we denote the multiphase system again by $(X_n)_{n\geq 0}$ now acting on S instead of $\{1, 2, \ldots, N+1\}$, i.e. X_n has the form $(i, k) \in S$ for all $n \geq 0$. We encode the particle's location in the first and the two phases in the second entry of our state space variable. The entrance is now in cell $(1, 0)$ and the label $(N+1, 0)$ is attached to the absorbing exit. The transition probabilities are given as:

$$\begin{aligned} p_{(i,k)(i-1,k)} &= \delta_i^{(k)}, \\ p_{(i,k)(i+1,k)} &= \beta_i^{(k)}, \\ p_{(i,k)(i,k)} &= \alpha_i^{(k)}, \\ p_{(i,k)(i,|k-1|)} &= \lambda_i^{(k)} \end{aligned}$$

and demand

$$\begin{aligned} \alpha_{N+1}^{(0)} &= 1, \\ \delta_i^{(k)} + \beta_i^{(k)} + \alpha_i^{(k)} + \lambda_i^{(k)} &= 1, \end{aligned}$$

for $k \in \{0, 1\}$, $1 \leq i \leq N$.

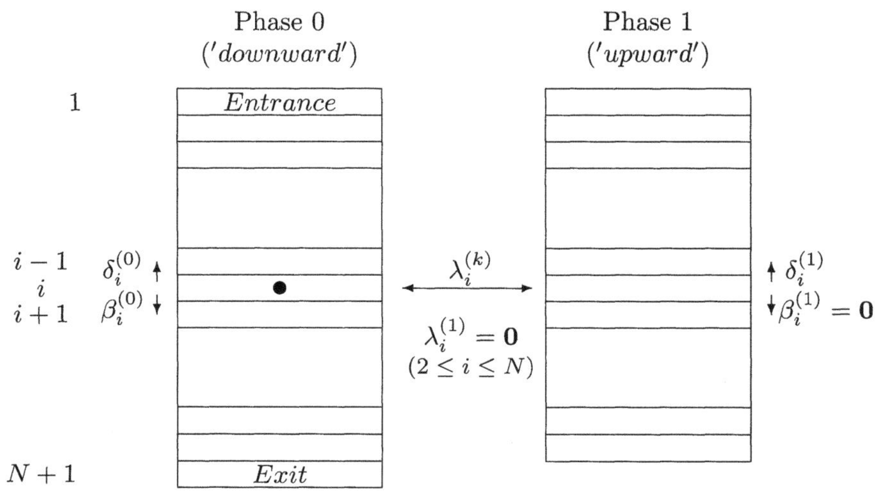

Figure 8.2: Multiphase system for a fluidized bed with arrows indicating all possible transition paths for the particle located in cell $(i, 0)$.

Phase 0 models the downward flow as a usual drift and diffusion and phase 1 models the upward drift in the wake of a rising gas bubble by setting

$$\beta_i^{(1)} = 0$$

for $1 \leq i \leq N$. A value greater than zero for the parameters $\beta_i^{(1)}$ would model a drifting and diffusing flow in phase 1 qualitatively similar to that in phase 0, but quantitatively different either upward ($\beta_i^{(1)} < \delta_i^{(1)}$) or downward ($\beta_i^{(1)} > \delta_i^{(1)}$) in each cell i. This behavior can be found in beds containing two different kinds of flow regimes, e.g. because of gulf streaming.

The change between the two phases, i.e. entering/leaving the wake of a gas bubble, is governed by $\lambda_i^{(k)}$. Setting

$$\lambda_i^{(1)} = 0$$

for $2 \leq i \leq N$ models that the upward phase will only be left at the top of the reactor. It is also possible to allow a change of phase in some specified cells or simply in all cells to model that a particle rising in the wake of a gas bubble could leave this bubble in these specified locations. This would correspond to a fraction of material that is rising in the wake of gas bubbles leaving these bubbles to start the standard downward drift and diffusion again. An example where this kind of modeling approach could gainfully be employed are e.g. fluidized bed reactors with baffles .

Summing up we obtain as possible transitions :

- Staying in the same cell, i.e. maintaining location and phase.
- Moving one cell upward, i.e. changing the location but not the phase.
- Moving one cell downward, i.e. changing the location but not the phase.
- Changing the phase, i.e. maintaining the location but switching the phase.

The model is visualized in Figure 8.2.

It should be emphasized that this multiphase system is a Markov chain with all its advantages in contrast to the particle's location only which is the projection of the multiphase system on its first coordinate/entry and not Markovian.

Dehling, Hoffmann and Stuut [39] successfully formulated a Markov chain model for particle movement in a fluidized bed. It introduced a parameter to describe a particle's rise in the wake of a gas bubble. In consequence an instant jump of the particle to the top of the bed was permitted. A state space just encoding the height of the particle in the bed was used still letting it be possible to formulate the process with a Markovian character.

In contrast the present extended model with two phases cannot be Markovian while only keeping information about the height of the particle in the fluidized bed in the state space. The implementation of a particular phase describing a particle's rise in the wake of a gas bubble causes the one-step-transition probabilities to not only depend on the particle's height in the fluidized bed but also on its current phase. To derive these transition probabilities knowledge about the current phase of the particle is therefore indispensable. Focusing only on the particle's height in the fluidized bed would deny the formulation of such a two phase model as a Markov process. However, the power of the theory of Markov processes has been shown before and should be used to its fullest. Thus it is essential to guarantee that the model is Markovian. This is ensured by enlarging the state space such that it encodes all necessary information to completely determine the one-step-transition probabilities.

Enlarging the state space turns out to be a fruitful detours to gain the sought information about the behavior of the examined system. An added advantage is that extra information can be obtained using the multiphase system approach, for instance it keeps track of the particle in each phase, i.e. the expected residence time of the particle in each phase can also be derived.

8.2 Gulf Streaming in Fluidized beds

We will now use the principle of stochastic modeling of multiphase systems on the case of a fluidized bed with gulf streaming. We begin with outlining a series of experiments elucidating the physical phenomena taking place in fluidized beds with gulf streaming, and then continue to formulate a stochastic model for this system.

8.2.1 Physical Phenomena in Fluidized Beds with Gulf Streaming—Experimental Evidence

This section gives an account of experiments carried out in a positron emission tomography (PET) camera, and reported in three recent articles [37, 38, 61] by some of the authors of this book.

Technique and Analysis

The PET technique is based on detecting two photons emitted in opposite directions as a result of the decay of a radioactive labeler. The decay of the labeller, here ^{18}F, leads to the emission of a positron and a neutrino. The positron travels a couple of millimeters through the system before annihilating with an electron (see Figure 8.3). This annihilation normally results in the emission of two gamma photons of 511 keV, which are emitted back-to-back so that energy and linear momentum are conserved.

The emitted photons are detected by cylindrically arranged scintillating crystal sensors. For these experiments an ECAT EXACT HR+ scanner was used, and the crystal sensors consisted of 72 blocks, each arrayed in 8×8 detectors. In depth there are 4 such rings of blocks, giving 4×72×64 detectors in total. A sketch of the arrangement (with fewer blocks) is shown in Figure 8.4. The dimensions of the detectors is 4.29×4.05 mm in the angular and z-direction, respectively. There is some space between the detectors; the diameter of the detector ring is 83 cm and the depth 15 cm.

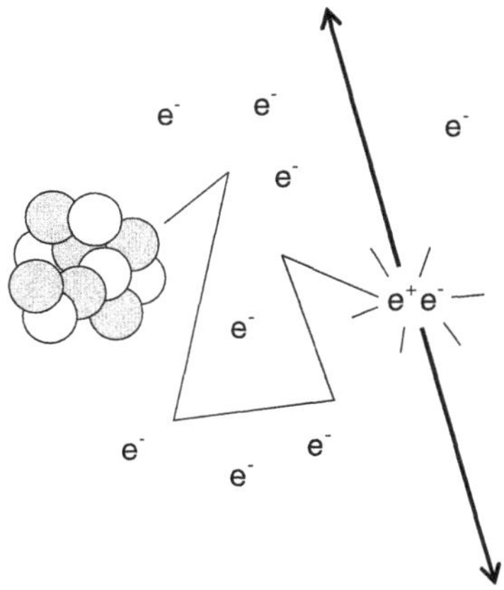

Figure 8.3: Sketch showing the emission of a positron and its annihilation with an electron

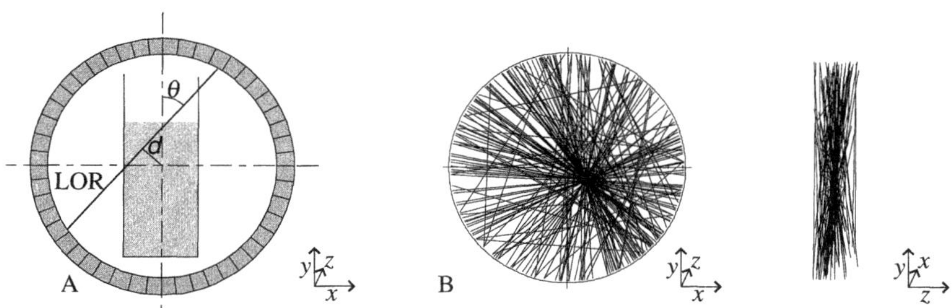

Figure 8.4: A: Sketch of the fluidized bed within the cylindrical detection zone of the camera, detector blocks and an LOR with its defining parameters are sketched. B: about 150 LORs out of the 500 detected within one millisecond

If two detections are within a narrow time window, they are adjudged to have emanated from the same annihilation event, and a 'line of response' (LOR) is drawn between the two detectors. The position of the LOR is stored as:

- its distance (in number of detectors) to the camera center, d in Figure 8.4,
- its angle (in number of detectors) with the vertical, θ in Figure 8.4 and
- the distance (in number of detectors) of each of the two detectors from the back of the detection volume, i.e. normal to the paper in Figure 8.4, and not shown.

This information is stored in two ways.

One way is event-by-event storage of the distance and angle data for the individual LORs. This is useful for tracking a single tracer particle, determining its position at successive points in time by cross-triangulation using LORs within a narrow time interval around the point of time in question.

The second way of storing data is to accumulate data for LOR distances and angles during a given period in time, and store them as a 2-dimensional array in the form of a "sinogram". Algorithms can then reconstruct the shape of the volume from which the events originated during the given period in time from the sinogram. In this research sinograms were obtained within fixed periods of time at the finest time resolution the software allowed: one sinogram per second. This second way of storing data is mainly useful for studying pulses of marked particles.

For more details about the data storage and retrieval, see the thesis of Nichols [99] and the paper of Hoffmann et al., [61].

For the tracking of one particle it was decided to determine one particle position per millisecond (ms). Bad LORs[1] were first eliminated, and the particle position in the x-y plane was then determined by averaging the cut-points between the rest. Computing power was a limiting factor for the analysis; the number of operations for computing particle positions increase with the square of the number of LORs included in the analysis. Writing a program in a compiled language (PASCAL), made it possible to include

[1]Some bad LORs can be seen in Figure 8.4. These may be caused by the detectors detecting one end each of two different LORs within a narrow time window, erroneously assigning them to one event. Detecting only one end of an LOR may occur if, for example, the other end falls outside the area covered by the detectors or if the other end passes through the detectors without triggering them.

100 LORs per ms in an analysis of an experiment lasting about 15 minutes using around 2 hours on a standard PC for the computations. The most efficient algorithm was one of 'zeroing in' on the average of the best cut-points by successive elimination of less precise cut-points, i.e. narrowing the spatial window containing the cut-points used for the averaging. The optimal final window size was around 10 mm. Like the crystal coordinates, the position data were generated in a Cartesian coordinate system, which is convenient for viewing projections on the coordinate planes. Once the x-y position was found, the z-position was determined by a similar procedure in the y-z plane, using only the cut-points contributing to the x-y average.

For the reconstruction of images of pulses of radioactive particles from the sinograms, the camera software was used, with some post-processing added. The reconstructed images are stored by the camera software in so-called "image files". In the case of the ECAT EXACT HR+ scanner an image is stored as radiation intensity in 128 by 128 by 63 voxels (three-dimensional pixels of 5.148 by 5.148 by 2.381 mm). For this research these images were constructed once per second and converted into files of one-byte numbers representing one voxel. Software was developed in-house to further analyze these files, making it possible to analyze the intensity data in the three coordinate planes. More details are given in Dechsiri et al. [38].

Experimental Materials and Apparatus

Two cylindrically shaped glass tubes of diameters 15 and 10 cm, respectively, and both of height 35 cm were used as fluidized bed reactor. Both near the top and near the bottom of each reactor porous sintered plates covered the cross-section. The bottom plate was of a fine grade to act as a distributor for the fluidizing gas, the top one was somewhat coarser and was installed only to prevent particles escaping into the hospital environment.

Two types of particles were used: a macroporous anion exchange resin and a catalyst powder for fluidized catalytic cracking (FCC). Their physical properties are given in Table 8.1.

Table 8.1: Physical properties of the particles used

	Lewatit MP500	FCC-catalyst
Average particle size [μm]	470	79.5
Density [kg/m^3]	1060	1464
Sphericity [-]	1	0.9
U_{mf} calc'd [132] [m/s]	0.075	0.45
U_{mf} meas'd [m/s]	0.116	0.004

Ion exchange resins work by ions residing on the resin being replaced by ions in solution. They can be chosen to be selective for F-ions, and the process of marking these resins is very simple. The strongly basic Lewatit MP500 resin was chosen, after initial tests showed it to be sufficiently effective in adsorbing Fluoride ions even when only a single tracer particle is used. The FCC-catalyst was used for experiments wherein a pulse of radioactive particles were traced. Both types of tracer particles were porous.

Results

Figure 8.5 shows results for the height of a single radioactive particle in a bed of the anion exchange resin Lewatit MP500 fluidized at a superficial velocity of $U = 0.130$ m/s. Plot A shows the particle position during the still period before the bed is fluidized and plot B the particle position in the fluidized bed.

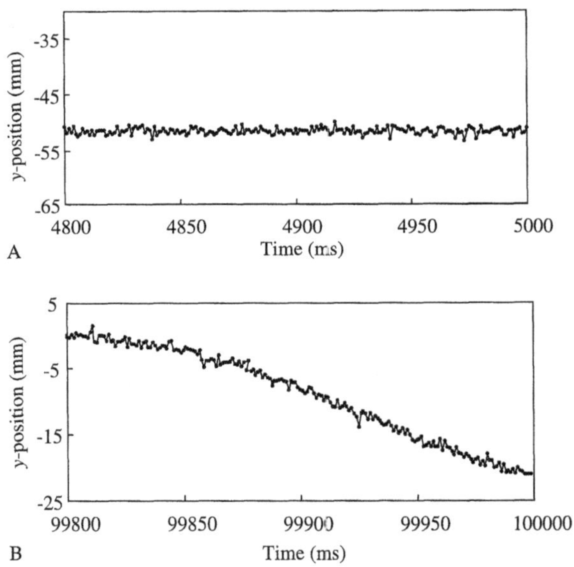

Figure 8.5: Particle y-position (height in the fluidized bed) as a function of time. A: still bed B: fluidized bed

There are a number of reasons for spatial scatter in the LORs:

1. The finite size of the detectors limits spatial resolution, and therefore the LOR precision.
2. The positron travels a few mm before annihilating with an electron.
3. The emission of photons may not be exactly collinear, i.e. 180° apart, this depends on the combined momentum of the electron-positron pair just before the annihilation.
4. The tracer particle itself has finite dimensions.

The scatter mentioned under point 1. depends on the particle's position in the field of view.

One way of estimating the variance in the results due to experimental scatter only is the 'variate method' (Kendall [75]), which will extract the scatter due to the purely random element of the millisecond particle positions in the following way. Say the elements, ε_t, of a discrete time series have the mean zero and a constant variance. The variance in the rth difference:

$$\Delta^r \varepsilon_t = \varepsilon_{t+r} - \binom{r}{1}\varepsilon_{t+r-1} + \binom{r}{2}\varepsilon_{t+r-2} - \ldots \pm (-1)^r \varepsilon_t$$

is:

$$\mathrm{var}(\Delta^r \varepsilon_t) = \binom{2r}{r}\mathrm{var}(\varepsilon).$$

Thus, if differences of order high enough to eliminate the trend in the data are taken, the expression should give an estimate of the scatter due to the purely random element in the data. In other words: when taking successively higher order differences, this expression should settle down to a constant, which is an estimate of the scatter due to the random element, once differences are of high enough order for the trend to be eliminated. Again this can be done for the still bed (no trend, so zeroth order differences should be enough) and the fluidized bed, and a comparison can be made.

Applying the variate methods to both moving and still beds showed that the scatter due to random motion is the same in the two [61]. Also it turned out that the variate method gave an approximately constant value for variance of the differences right from the first order difference, probably due to the trend in the results being very weak in comparison with the scatter.

To obtain the full information from the results it is necessary to determine over how long a time interval the particle positions be averaged without loosing information about the real particle movement.

An answer to this question can be given by realizing that replacing a group of equally spaced data points in a time-series with their arithmetic average at the middle time step is equivalent to fitting them with a straight line. If we therefore do this using successively larger groups (or time intervals) for our linear fits, we should find a point where the estimated residual standard deviation starts to increase above the value obtained from the variate method, signifying that linear fits are no longer enough to describe the trend in the data *i.e.* the real particle motion.

The result of doing this on a limited time interval is shown in Figure 8.6. When only a few points are used for each linear fit, the fitted line segments are somewhat able to follow coincidental 'trends' in the data, explaining why the line-fitting method gives a slightly low estimate for the standard deviation near the origin. When more than 60 points are used for each fit, the estimate of the standard deviation from the line fitting method becomes too high, showing that the trend, or aspects of the particle motion, begins to be missed. The plot indicates that a true picture of the particle motion is gained if 10-50 points are replaced by their arithmetic average at the central time step.

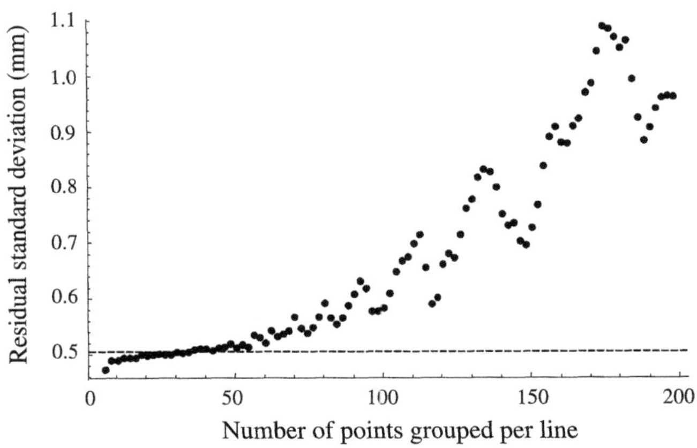

Figure 8.6: Residual standard deviation using the linear fit method as a function of the size of the number of millisecond particle positions used for each linear fit. The estimate for the standard deviation due to the random element from the variate method is indicated as a broken line

The particle motion resulting from averaging the millisecond data in this way are shown in Figure 8.7.

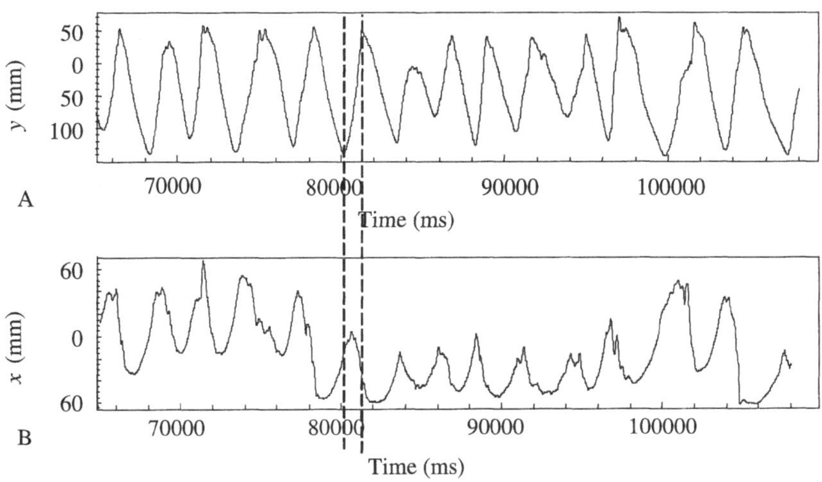

Figure 8.7: Time-series of the tracer particle's y-position (height in bed) and x-position

It is clear from this that the particles generally rise faster than they descend, consistent with quick rise associated with the bubble phase and slower descent in the bulk, as discussed in Section 3.2.1. However, when computing the expected descent velocity using the relations given in that section, it turns out to be much smaller than that seen in the figure. This indicates that there is far more vertical transport than can be accounted for by transport in bubble wakes. Figure 8.8 shows 3-D plots of particle tracks seen from two angles (A and B) and a plot showing the particle velocities over a cross-section in the middle of the bed (C).

This figure clearly shows that the particle circulates in the bed, rising rapidly—but with a large spread in the velocities—in one part of the bed and descending more slowly in another part. This indicates "gulf streaming": a region of high bubble activity exists wherein the particles are transported up, not only in bubble wakes but also in the bulk material between the bubbles. In the region of low, or no, bubble activity, the particles descend. This effect is probably due to less-than-optimal gas distribution in this laboratory bed, but gulf streaming is known to exist almost ubiquitously in large industrial beds, if not due to inferior gas distribution over the cross-section, then due to the fluidization bubbles concentrating near

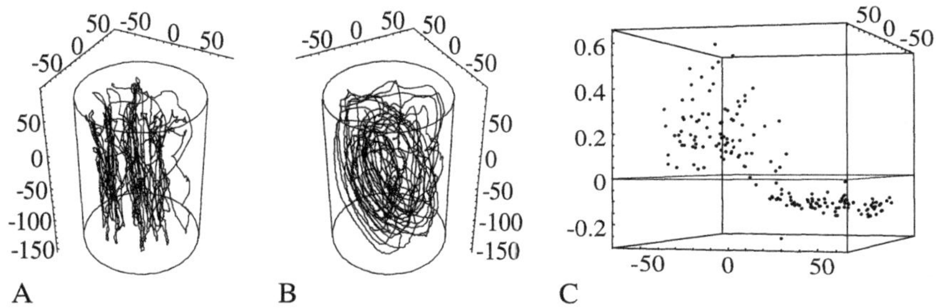

Figure 8.8: A and B: 3-D plots of particle tracks in the fluidized bed seen from two different angles C: The particle velocities over a cross-section in the middle of the bed

the axis as a consequence of coalescence.

Interpretation

We assume that the bed cross-section can be split in two regions: one wherein there is a strong bubble activity and a generally upwardly directed particle flow, either in the wakes of fluidization bubbles, or in the material between the bubbles and another wherein the particle flow is downward and the bubble activity is low or absent. Figure 8.9, wherein the start and endpoints of upward particle paths are shown support this view, and seems to indicate that the upward flow takes up about 1/3 of the bed cross-section.

Figure 8.10 shows a frequency plot for the particle velocities shown in Figure 8.8C. Two local maxima at about -0.125 m/s and +0.2 m/s are clear, and a few times the tracer particle can be seen to have passed the middle with a velocity of 0.5-0.6 m/s. This leads us to assume the velocity of descent, $u_{s,d}$ is about 0.125 m/s, and the velocity of ascent in the bulk phase between the fluidization bubbles, $u_{s,u}$ is 0.2 m/s while the velocity of the bubble wakes, which follow the fluidization bubbles, u_b, may be 0.5–0.6 m/s. We will verify whether these values are mutually consistent, and consistent with the empirical equations given in Chapter 3.

We will call the fraction of cross-sectional area taken up by upward flow, which, as mentioned, we estimate to be 1/3, A_u, and the fraction with downward flow A_d, we call the corresponding superficial gas velocities relative to an observer U_u and U_d, respectively.

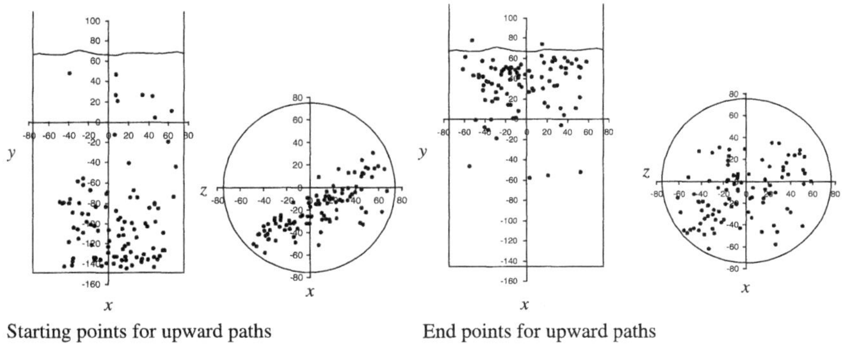

Figure 8.9: Start and endpoints for upward particle paths in the bed. Figures are redrawn from [37], where they, however, unfortunately have been interchanged

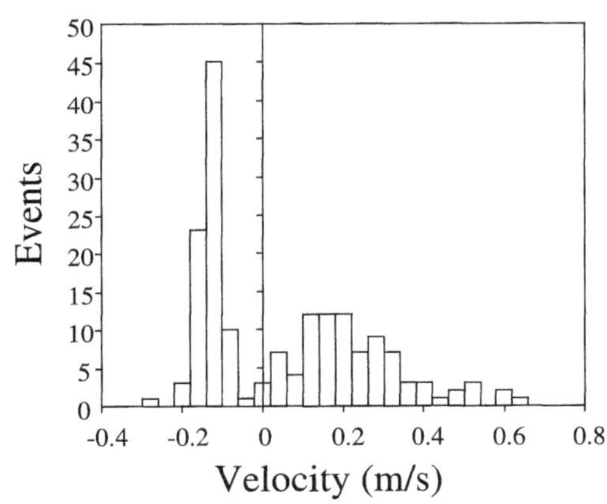

Figure 8.10: Frequency plot for the velocities shown in Figure 8.8C

In accordance with the two-phase theory, Equation (3.1), we assume that the superficial velocity of the interstitial gas relative to the fluidized solids remains at the minimum fluidization velocity, $U_{mf} = 0.116$ m/s everywhere in the bed, also if it is bubbling, giving a real gas velocity relative to the particles of

$$\frac{U_{mf}}{\varepsilon_{mf}} = \frac{0.116}{0.42} = 0.276 \text{ m/s}.$$

As is often done in this type of analysis, we will neglect the small cross-sectional area taken up by the fluidization bubbles.

In the down-flowing part of the bed, which we assume to be bubble-free and in which the particles flow down with a velocity of about 0.125 m/s, the gas velocity relative to an observer is then $0.276 - 0.125 = 0.151$ m/s and thus this gas will have a superficial velocity relative to an observer of $U_d = 0.151 \times \varepsilon_{mf} = 0.151 \times 0.42 = 0.0635$ m/s.

For continuity (volume conservation):

$$A_u U_u + A_d U_d = U \;\Rightarrow\; U_u = \frac{1}{A_u}(U - A_d U_d)$$
$$= 3 \times (0.13 - 0.0635\frac{2}{3}) = 0.263 \text{ m/s},$$

which means that *if the up-flowing part of the bed did not bubble, but the voidage fraction in the particles remained* ε_{mf}, the real velocity in that part relative to an observer would be:

$$\frac{U_u}{\varepsilon_{mf}} = \frac{0.263}{0.42} = 0.626 \text{ m/s},$$

making the real gas velocity relative to the upward moving particles between the bubbles $0.626 - 0.2 = 0.426$ m/s.

The upward flow region can be viewed as an upwardly moving fluidized bed fluidized with a superficial gas velocity, $U_{u,r}$ of $0.426 \times \varepsilon_{mf} = 0.426 \times 0.42 = 0.179$ m/s, and this will, according to the two-phase theory (3.1), bubble and the superficial velocity of the gas flowing in the bubble voids will be:

$$(U_{u,r} - U_{mf}) = 0.179 - 0.116 = 0.063 \text{ m/s}.$$

We can now calculate the mean bubble size in this section by integrating the Mori and Wen equation for bubble size as a function of height in the bed, and dividing by the apparent total height of the bed[2]. Using the physical value for D_{bed} of 0.15 m, this gives a mean bubble size of 0.0286 m in the upflow region.

According to Equation (3.3), bubbles of this size will have a wake fraction, f_w, of 0.262.

To obtain the rise velocity of fluidization bubbles, we may use the semi-empirical relation by Rowe [107]:

$$u_b = 0.711\sqrt{gD_b}, \tag{8.1}$$

where g is the gravitational acceleration. Inserting $D_b = 0.0286$ m gives a bubble velocity of 0.377 m/s.

Thus particles in the wakes of fluidization bubbles should rise with a velocity of $0.2 + 0.377 = 0.577$ m/s relative to an observer. We see that this last result agrees very well with the value of 0.5–0.6 m/s we surmized from the experimental results.

The volumetric flow of particles (with their associated void space) per unit area in the bubble wake phase relative to the upward flowing particles between the bubbles is:

$$\frac{Q_{w,r}}{A_u} = (U_{u,r} - U_{mf})\frac{f_w}{1 - f_w} = (0.179 - 0.116)\frac{0.262}{1 - 0.262} = 0.0224 \text{ m/s}.$$

Relative to an observer the bubble move with a velocity of 0.577 m/s rather than 0.377 m/s, so that the total flow in the bubble wake phase per unit area becomes: $Q_w/A_u = 0.0224 \times (0.577/0.377) = 0.0343$ m/s.

We find that the nett volumetric particle flow in the bed (which should be zero) becomes:

$$-A_d u_{s,d} + (u_{s,u} + \frac{Q_w}{A_u})A_u = -\frac{2}{3}0.125 + (0.2 + 0.0343)\frac{1}{3} = -0.0052 \text{ m/s},$$

which is very low.

Thus the experimental observations are mutually consistent, and consistent with the empirical equations given for a bubbling fluidized bed in Chapter 3. This supports the interpretation of the experimental results given above, and this interpretation will form the basis of the model formulated below.

[2]For this calculation the physical bed height is reduced to the apparent bed height experienced by a bubble due to the up-flow of the solids between the bubbles by multiplying the physical bed height by $1 - u_{s,u}/u_b \approx 1 - 0.2/0.55$,. In these rough calculations, we do not iterate for the precise value of u_b.

8.3 Extension of the Model to include Gulf Streaming

The aim of this section is to extend the above model such that exchange of material between bubble and bulk flow is incorporated and the phenomenon of gulf streaming is captured. Exchange of material between bubble and bulk flow is always possible and usually an important issue if e.g. binary mixtures of particles with different densities are present. Then the particles with higher density segregate downward and form the major part of wake material since the gas bubbles originate at the bottom of the reactor. When these particles move upward in the wake of a rising gas bubble they can, probably driven by the difference in average density between the wake and the surrounding bulk, leave the bubble and enter the bulk phase. Also baffles may cause wake material to enter the bulk flow by knocking out the wakes from the rising bubbles [19, 33, 63].

Gulf streaming has its cause in a cross-sectionally non-uniform bubble flow inducing an extra upward flow in one part of the bed and a corresponding downward flow somewhere else in the bed. Consequences are a large amount of material from the bottom transported to the top and an enhanced velocity in the downward bulk flow. Gulf streaming is a common effect in fluidized beds especially large industrial ones and the dominating axial mixing mechanism when compared to the mixing by the flow of wake material in bubbles. In the previous section, we found that, of the total solids upflow per unit area of about 0.263 m/s, only about 0.0224 m/s was due to flow in the bubble wakes. [89, 91, 133].

From now on we consider a closed reactor without an exit and delete the state $(N+1, 0)$ from the state space of the multiphase system. To incorporate the possible exchange of wake material and bulk flow we set the parameter that governs this exchange $\lambda_i^{(k)} > 0$. Then the fraction of wake material in a rising gas bubble that leaves the gas bubble and joins the bulk flow at location i which is the interpretation of $\lambda_i^{(1)}$ is greater than zero.

Including gulf streaming needs a bit more effort because we have to enlarge the state space again. But we do not have to introduce a whole new type of phase. Enlarging the state space to

$$S = \{1, 2, \ldots, N\} \times \{0, 1, 2\}$$

is sufficient. Then the first entry between 1 and N still encodes the particle's location and the second entry encodes its phase now either

- Phase 0: downward bulk flow with minimal dispersion or

- Phase 1: upward flow in the wake of a rising gas bubble with minimal dispersion or
- Phase 2: upward flow between the bubbles via gulf streaming with dispersion.

Phase 0 models movement in the part of the reactor without bubble activity which therefore exhibits only minimal dispersion. A higher value of dispersion is present in phase 2 due to disturbance of the upgoing stream by the fluidization bubbles 'carrying' it.

The corresponding transition probabilities are now given as

$$\begin{aligned} p_{(i,k)(i-1,k)} &= \delta_i^{(k)}, \\ p_{(i,k)(i+1,k)} &= \beta_i^{(k)}, \\ p_{(i,k)(i,l)} &= \lambda_i^{k,l} \end{aligned}$$

and we require

$$\delta_i^{(k)} + \beta_i^{(k)} + \sum_{l=0}^{2} \lambda_i^{k,l} = 1 \tag{8.2}$$

for $k, l \in \{0, 1, 2\}$, $1 \leq i \leq N$. The probability to stay in a given cell $p_{(i,k)(i,k)}$ is here denoted by $\lambda_i^{k,k}$ replacing the parameter α found in Chapter 3.

The possible transitions are thus the same as in the model in Section 8.1 with one additional phase to change to in step four below:

1. Staying in the same cell, i.e. maintaining location and phase.
2. Moving one cell upward, i.e. changing the location but not the phase.
3. Moving one cell downward, i.e. changing the location but not the phase.
4. Changing the phase, i.e. maintaining the location but changing the phase.

We picture the situation in Figure 8.11.

Now the model for the bubbling fluidized bed with exchange between wake and bulk flow and gulf streaming is set up. To finish it the transition probabilities have to be assigned in such a way that the physical processes are correctly modeled. This is usually a hard task especially concerning the parameters $\lambda_i^{k,l}$ governing the interchange between the phases. The physical velocities and dispersions have to be converted to dimensionless

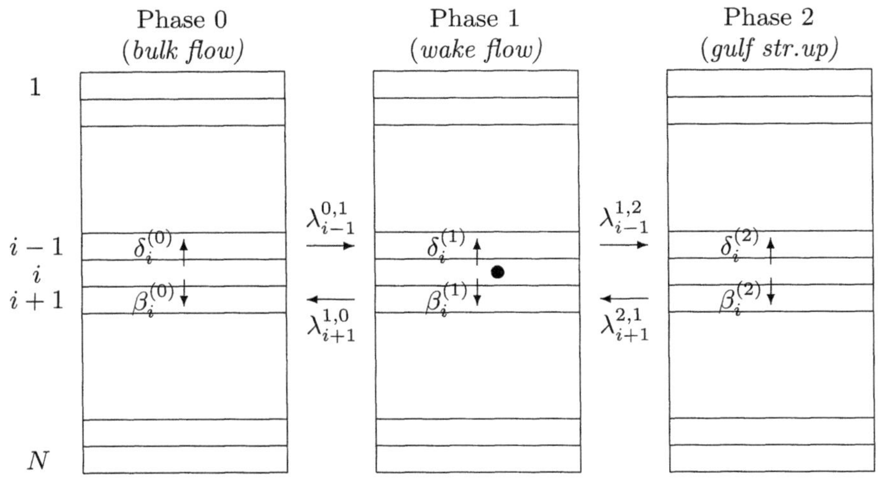

Figure 8.11: Extended multiphase system for a fluidized bed with arrows indicating some possible transition paths and the particle located in cell $(i, 1)$.

ones. For given spatial and temporal discretizations Δ and ε relating to physical units as follows

$$\Delta = M \times \mathrm{m} \text{ and } \varepsilon = S \times \mathrm{s}.$$

with some positive numbers M and S the conversion formulas are given as

$$\begin{aligned} v\frac{\Delta}{\varepsilon} &= v\frac{M}{S}\frac{\mathrm{m}}{\mathrm{s}} &= \tilde{v} \text{ (in m/s)}, \\ D\frac{\Delta^2}{\varepsilon} &= D\frac{M^2}{S}\frac{\mathrm{m}^2}{\mathrm{s}} &= \tilde{D} \text{ (in m}^2\text{/s)}, \end{aligned} \tag{8.3}$$

with dimensionless velocity v and dispersion D and physical velocity $\tilde{v}$ and dispersion $\tilde{D}$. This yields a mean displacement of

$$(\Delta \times \beta_i^{(k)} - \Delta \times \delta_i^{(k)})\frac{1}{\varepsilon} = (\beta_i^{(k)} - \delta_i^{(k)})\frac{\Delta}{\varepsilon}$$

and a mean squared displacement of

$$(\Delta^2 \times \beta_i^{(k)} + \Delta^2 \times \delta_i^{(k)})\frac{1}{\varepsilon} = (\beta_i^{(k)} + \delta_i^{(k)})\frac{\Delta^2}{\varepsilon}$$

for a given cell (i, k). Dispersion enters into the model by equalizing the mean squared displacement and two times the dispersion coefficient. This is a reasonable simplification since the physical dispersion, usually seen as the variance which is given by mean displacement squared subtracted from mean squared displacement, is approximated by its dominating term the mean squared displacement. Therefore true plug flow can only be modeled in one way namely by setting the corresponding transition probability δ or β equal to one and choosing the time and space discretization such that $\frac{M}{S} = \tilde{v}\frac{\mathrm{s}}{\mathrm{m}}$.

We consider the transition probabilities for each phase first and those between the phases thereafter.

- *Bulk flow:* The necessary extra conditions for the bulk flow need to minimize dispersion and ensure a downward flow. The dimensionless velocity at each location i is given as $\beta_i^{(0)} - \delta_i^{(0)}$ and the dimensionless quadratic displacement as $\beta_i^{(0)} + \delta_i^{(0)}$. Setting the flow parameter $\delta_i^{(0)}$ to be zero allows no upward flow in the bulk and minimizes dispersion.

- *Wake flow:* As for the bulk flow the dimensionless velocity at each location i is given by $\beta_i^{(1)} - \delta_i^{(1)}$ and the dimensionless quadratic displacement as $\beta_i^{(1)} + \delta_i^{(1)}$. Analogous to the bulk flow the wake flow has no downward movement restraining the wake flow parameter $\beta_i^{(1)}$ to be zero. This also minimizes dispersion as above.

- *Gulf streaming upward:* Again the dimensionless velocity at each location i is given by $\beta_i^{(2)} - \delta_i^{(2)}$ and the dimensionless quadratic displacement as $\beta_i^{(2)} + \delta_i^{(2)}$. Upward gulf streaming shows dispersion, since this is the region through which most of the bubbles flow. Thus we impose that $\beta_i^{(2)} = D_i + \frac{1}{2}v_i$ and $\delta_i^{(2)} = D_i - \frac{1}{2}v_i$ to generate a dispersion with dimensionless dispersion coefficient of $2D_i$ and a dimensionless velocity of v_i at location i.

We note that the choice of $\beta_i^{(2)}$ and $\delta_i^{(2)}$ in the gulf streaming upward above is the only possibility to yield a dimensionless mean displacement of

$$\beta_i^{(2)} - \delta_i^{(2)} = v_i$$

and a dimensionless mean squared displacement of

$$\beta_i^{(2)} + \delta_i^{(2)} = 2D_i$$

for a cell $(i, 2)$ conditioned on maintaining the phase. Between the phases the following restraints have to be made:

1. Flow between the gulf streaming phase and all other phases almost only takes place in the upper and lower regions of the reactor. Thus the parameters $\lambda_i^{k,l}$ have to be very small if $k = 2$ or $l = 2$ and $i \in \{i_0 + 1, \ldots, i_1 - 1\}$ for some $1 \leq i_0 \leq i_1 \leq N$. Then $\{1, 2, \ldots, i_0\}$ constitutes the upper and $\{i_1, \ldots, N\}$ the lower part of the reactor where most transitions involving the gulf streaming phase take place.

2. No other restrictions on the transition probabilities $\lambda_i^{k,l}$ are made to allow an exchange between wake material and bulk flow at each location.

We assemble the demands above for an overview.

	up/down	transition to wake/bulk ($l \in \{0, 1\}$)	transition to gulf str.
Bulk flow	$\delta_i^{(0)} = 0$	–	$\lambda_i^{0,2} \sim 0$ for $i_0 < i < i_1$
Wake flow	$\beta_i^{(1)} = 0$	–	$\lambda_i^{1,2} \sim 0$ for $i_0 < i < i_1$
Gulf str.up	$\beta_i^{(2)} = D_i + \frac{1}{2}v_i$ $\delta_i^{(2)} = D_i - \frac{1}{2}v_i$	$\lambda_i^{2,l} \sim 0$ for $i_0 < i < i_1$	–

Here the symbol ~ 0 is used in the sense of 'very small' or 'next to 0'.

When the particles in the reactor are incompressible no mass can accumulate in any given cell such that that the mass balance equation

$$\sum_{i=1;k\in\{0,1,2\}}^{N} p_{(i,k),(j,l)} = \delta_{j+1}^{(l)} + \beta_{j-1}^{(l)} + \sum_{k=0}^{2} \lambda_j^{k,l} = 1 \tag{8.4}$$

has to hold for all $(j, l) \in \{1, \ldots, N\} \times \{0, 1, 2\}$. This is in the spirit of Section 4.3. Here it should be pointed out that the mass balance equation above rests on the assumption of all cells having the same volume, i.e. in the present case even for different phases $k \in 0, 1, 2$. This should be taken into account when deriving the transition probabilities $\beta_i^{(k)}, \delta_i^{(k)}$ within the phases.

8.4 Quantification of the Model Parameters

Having Chapter 2 in mind we have to assign the discretization size Δ, i.e. the height of one cell and then calculate the transition probabilities given the dimensionless velocity v_i, dimensionless dispersion D_i and phase transition rates $\lambda_i^{k,l}$ at each location i according to the following scheme.

- $\beta_i^{(k)} = D_i + \frac{1}{2}v_i$
- $\delta_i^{(k)} = D_i - \frac{1}{2}v_i$
- $\lambda_i^{k,l} = \bar{\lambda}_i^{k,l}$

$(1 \leq i \leq N; k, l \in \{0,1,2\}; k \neq l)$. The parameter $\bar{\lambda}_i^{k,l}$ denotes the dimensionless phase transition rates given as

$$\bar{\lambda}_i^{k,l} = \varepsilon \tilde{\lambda}_i^{k,l},$$

with the physical transition rates $\tilde{\lambda}_i^{k,l}$ given in 1/s. Thus a particle located at cell (i,k) has a probability of $\varepsilon\tilde{\lambda}_i^{k,l}$ to change from its current phase k to phase l during a time period of length ε. To calculate the dimensionless velocities and dimensionless dispersion coefficients v_i and D_i for the given discretization from those present $\tilde{v}_i$ and $\tilde{D}_i$ in m/s and m^2/s we need to convert time and space from seconds and meters into the discretization units: '$s \rightsquigarrow \varepsilon$' and '$m \rightsquigarrow \Delta$'. We obtain Δ as the quotient of height of the bed h and number of cells N

$$\Delta = \frac{h}{N}.$$

We refrain from choosing the timestep ε in the order of Δ^2 as in Chapter 2 because this would cause to increase calculation times drastically. The choice of $\varepsilon = \Delta$ still ensures a fine discretization and has the advantage that m/s converts to Δ/ε one to one. This results in each timestep having the length $\varepsilon = \frac{h}{N}$. Following Equations (8.3) the conversions between physical and dimensionless parameters are obtained as

$$v_i = \tilde{v}_i \text{ s/m}$$

and

$$D_i = \frac{N}{h}\tilde{D}_i \text{ s/m}.$$

From Section 8.2.1 and Figure 8.10 we obtain for the velocities $v^{(k)}$ corresponding to the three phases:

$$v^{(0)} = 0.125 \text{ m/s},$$

$$v^{(1)} = 0.577 \text{ m/s},$$

$$v^{(2)} = 0.2 \text{ m/s}.$$

The dispersion is minimal in phases 1 and 2 and for phase 3 Equation (3.8) gives

$$D^{(2)} = 1.7 \cdot 10^{-4} \quad \text{m}^2/\text{s}.$$

Based on the interpretation of Figure 8.10 given above the transition rates from the downward bulk flow to the upward gulf streaming flow and the upward flow in the bubble wakes have a ratio of 0.2 to 0.0343. This leads to

$$\frac{\sum_{i=i_1}^{N} \lambda_i^{0,1} \cdot \pi(i,0)}{\sum_{i=i_1}^{N} \lambda_i^{0,2} \cdot \pi(i,0)} = \frac{0.2}{0.0343}, \tag{8.5}$$

where $\pi(i,k)$ denotes the invariant distribution of $(X_n)_{n\geq 0}$ evaluated for cell (i,k). Assuming the parameters $\lambda_i^{0,1}$ and $\lambda_i^{0,2}$ to be constant for $i \in \{i_1, \ldots, N\}$ and using the fact that the two transfer probabilities must sum to 1 simplifies Equation (8.5) to

$$\begin{aligned} \lambda_N^{0,1} &= 0.854, \\ \lambda_N^{0,2} &= 0.146 \end{aligned}$$

for the bottom of the reactor and

$$\frac{\lambda_i^{0,1}}{\lambda_i^{0,2}} = 5.83$$

has still to hold for $i_1 \leq i \leq N-1$. It is worthwhile to note that Equation (8.5) becomes

$$\frac{\sum_{i=i_1}^{N} \lambda_i^{0,1}}{\sum_{i=i_1}^{N} \lambda_i^{0,2}} = \frac{0.2}{0.0343}$$

if π is the uniform distribution. Physical considerations namely the absence of segregation in the fluidized bed lead to the conclusion that the invariant (stationary) distribution is given by the uniform distribution. This justifies the simplification of Equation (8.5) using the uniform distribution as done above and gives another checkmark for deriving the model parameters.
The height of the bed in fluidized state is 0.22 m. We divide the bed in $N = 440$ equisized cells yielding $\Delta = 0.0005$ m and $\varepsilon = 0.0005$ s. The transition parameters for exchange between the phases $\lambda_i^{k,l}$ and the regions of exchange between gulf streaming and bulk determined by i_0 and i_1 are

difficult to determine and those not discussed above set as reasonable fit. The resulting transition probabilities now are.

$$
\begin{aligned}
\delta_i^{(0)} &= 0 && && \text{for all } 2 \le i \le N,\\
\delta_i^{(1)} &= v^{(1)}\ \tfrac{\mathrm{s}}{\mathrm{m}} &&= 0.577 && \text{for all } 2 \le i \le N,\\
\delta_i^{(2)} &= \left(\tfrac{N}{h} D^{(2)} - \tfrac{1}{2} v^{(2)}\right) \tfrac{\mathrm{s}}{\mathrm{m}} &&= 0.44 && \text{for all } 2 \le i \le N,\\
\beta_i^{(0)} &= v^{(0)}\ \tfrac{\mathrm{s}}{\mathrm{m}} &&= 0.125 && \text{for all } 1 \le i \le N-1,\\
\beta_i^{(1)} &= 0 && && \text{for all } 1 \le i \le N-1,\\
\beta_i^{(2)} &= \left(\tfrac{N}{h} D^{(2)} + \tfrac{1}{2} v^{(2)}\right) \tfrac{\mathrm{s}}{\mathrm{m}} &&= 0.24 && \text{for all } 1 \le i \le N-1,
\end{aligned}
$$

$$
\begin{aligned}
\lambda_i^{0,1} &= 0 && \text{for all } 2 \le i \le i_1 - 1,\\
\lambda_i^{0,1} &= 0.0007 && \text{for all } i_1 \le i \le N - 1,\\
\lambda_i^{0,2} &= 0.00075 && \text{for all } 2 \le i \le i_0,\\
\lambda_i^{0,2} &= 0.000078 && \text{for all } i_0 < i < i_1,\\
\lambda_i^{0,2} &= 0.004 && \text{for all } i_1 \le i \le N - 1,\\
\lambda_i^{1,0} &= 0 && \text{for all } 2 \le i \le N - 1,\\
\lambda_i^{1,2} &= 0 && \text{for all } 2 \le i \le N - 1,\\
\lambda_i^{2,0} &= 0.005 && \text{for all } 2 \le i \le i_0,\\
\lambda_i^{2,0} &= 0.000061 && \text{for all } i_0 < i < i_1,\\
\lambda_i^{2,0} &= 0 && \text{for all } i_1 \le i \le N - 1,\\
\lambda_i^{2,1} &= 0 && \text{for all } 2 \le i \le N - 1,
\end{aligned}
$$

$$
\begin{aligned}
\lambda_1^{2,0} &= 0.76,\\
\lambda_1^{1,0} &= 1,\\
\lambda_N^{0,1} &= 0.854,\\
\lambda_N^{0,2} &= 0.146
\end{aligned}
$$

with $i_0 = 80$ and $i_1 = 400$. All other transitions except for $\lambda_i^{k,k}$ are set to zero. The onestep return probabilities $\lambda_i^{k,k}$ are chosen in such a way that Equation (8.2) holds. The high value of N has its source in the high order of the difference of the dispersion coefficients and the velocities. It has to be assured that all transition probabilities are nonnegative, i.e. here $\frac{N}{h}D^{(2)} - \frac{1}{2}v^{(2)} \geq 0$.

8.5 Model Validation with Data

Using the model, data and parameters from Sections 8.3 and 8.4 a simulation of the path of one single particle starting at the bottom of the reactor, i.e. cell $(N, 0)$, is computed with MATLAB for 42 s. This simulation is compared to the experimental results from Dechsiri et al. [37]. Both paths are shown in Figure 8.12. Comparison of simulation and experiment

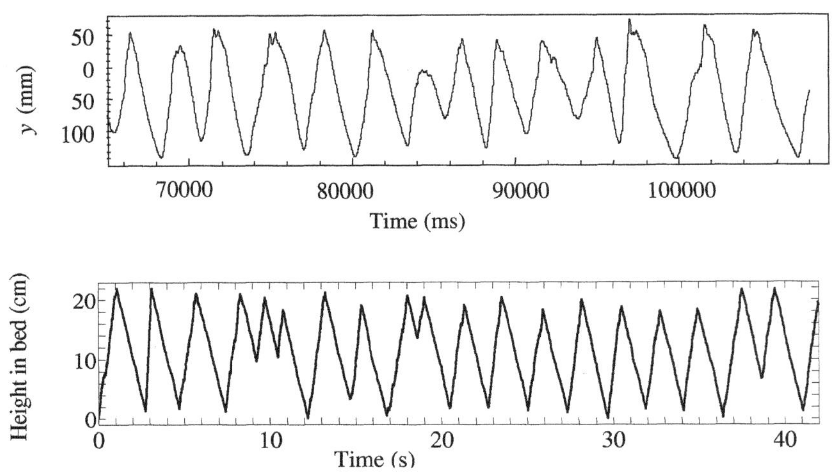

Figure 8.12: Time-series of the tracer particle's y-position (height in bed): experiment and simulation

provides the following.

1. The experimental results show 14 up and down passages of the particle. The simulation shows 16. This has to be expected since the parameters yield a downward velocity of 0.125 m/s and an averaged upward velocity of $(0.146 \times 0.577 + 0.854 \times 0.2)$ m/s $= 0.255$ m/s and

therefore a mean time for one up and down passage of $(0.22/0.125 + 0.22/0.255)$ s = 2.62 s. This leads to a mean number of up and down passages in 42 s of $42/2.62 = 16$ disregarding any phase changes.

2. The maxima of 13 of 14 passages in the experiment lie in the upper 4 cm of the reactor in agreement with the simulated path just as 11 of 14 minima being in the lower 2 cm of the reactor as 12 of 16 in the simulation.

3. There exist 4 phase transitions changing from a downward to an upward flow and 2 in the other directions in the middle of the reactor in the first picture. The simulated path exhibits 4 changes from downward to upward flow and only 1 change from up- to downward flow. All other phase changes take place in the upper 4 and lower 2 cm of the reactor. As in 2. consistency of experiment and simulation is valid.

4. The simulated path is much straighter and sharper around the extrema than the observed one. But this is no surprise since the velocities in the model are kept constant throughout the whole reactor. Obviously they should be adjusted at top and bottom regions to decrease in the neighborhoods of the extrema. Physically the particle has to change the sign of the velocity and therefore decrease its velocity to zero first.

5. In the upper region of the reactor fluctuating behavior of the particle's path is observed 4 out of 14 times. Even though the parameters in the model allowing just this behavior are set to values greater than zero the simulation lacks this behavior. Its cause could be chance since we examine a random simulation or two small values of the corresponding phase transition parameters.

Concluding, experimental and simulation results coincide to a high degree. The simulation captures all the essential features of the observed particle movement. Quantitatively simulation and experiment are close together although the velocities in the model are somewhat larger than those for the observed path. Likewise simulation and experiment behave qualitatively the same. The simulation still lacks minor characteristics as pointed out above in points 4. and 5. which had to be expected and/or chosen to neglect in the effort to present a comprehensible model. The most pressing issue concerning the presented model is the derivation of the phase transitions parameters.

8.5.1 Additional Qualitative Results

Calculation and simulation results using the approach from Section 8.3 with a starting layer in the middle of the reactor are shown below in Figures 8.13 and 8.14. Figure 8.13 shows the path of one single particle starting in the middle of the reactor for 10 seconds. It was generated using a simulation programmed in MATLAB. Figures 8.14 are of a different type. These give the particle's vertical distribution at different time steps calculated via raising the transition matrix to a power defining the output time and multiplying the starting distribution vector from the left. Here no simulation is involved only calculation. This not only gives more exact results but also a far lower computer-processing time involved than using simulations, e.g. Monte Carlo Markov chain methods. This effect took place when generating the pictures below.

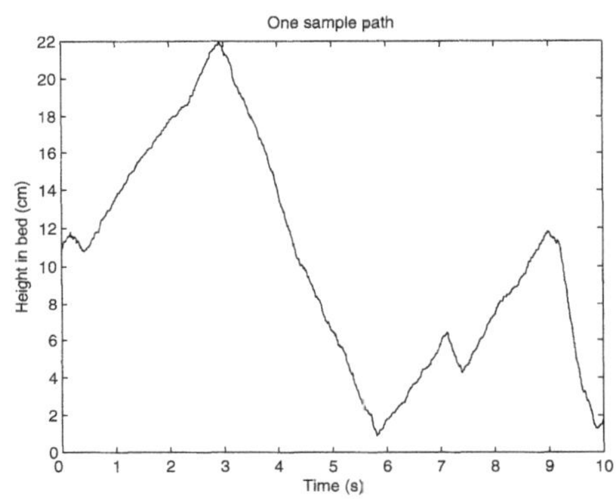

Figure 8.13: One path of the multiphase process

Figure 8.14 shows that the particles starting in the middle of the reactor move either up- or downward in three blocks. The by far largest block moves downward. Two smaller blocks move upward. The velocity of each block decreases with its size. This is in accordance with the model setup. The smallest block represents the particles in the bubble wake which move upward with a high velocity. The upward gulf streaming particles are found in the second smallest block. Most of the particles move slowly downward in the bulk flow constituting the largest block.

The velocities concur with those seen in Figure 8.15 (middle layer, 5-8). The downward bulk flow needs about 3 seconds to reach the bottom and moves linearly in between. The same picture is caught in the Figures 8.14

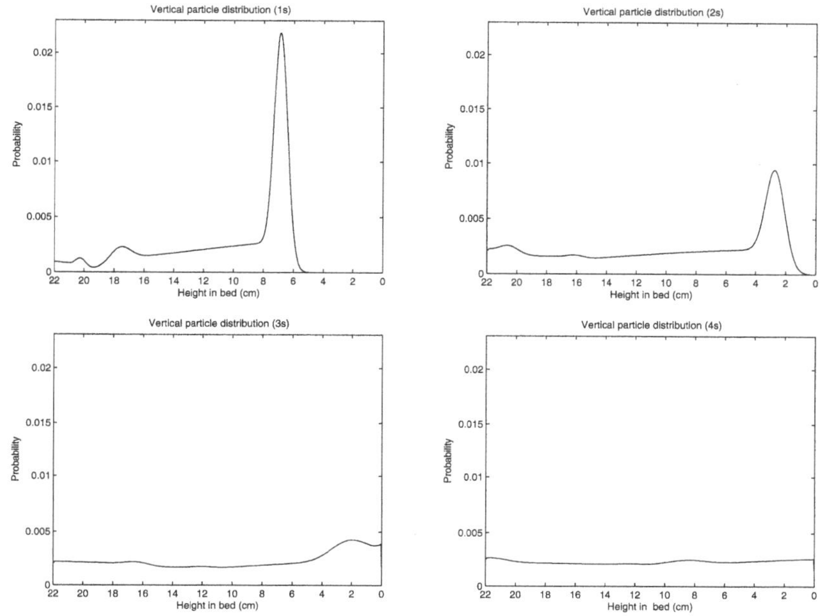

Figure 8.14: Vertical distribution of the particle for different times

for 1 to 3 seconds. After a time a bit longer than 1 second the upward flow reaches the top of the bed. This is both seen in the experimental and the modeling results.

The dispersion is much stronger in the upward going flow than in the downward going one, although the dispersion in the downward going flow increases over time. Again these features are seen in the experimental and the modeling results.

In spite of all the common features in experimental and model results mentioned above they differ slightly when comparing the pictures for 4 seconds (number 4 in Figure 8.14 and number 8 in Figure 8.15). The block at the bottom of the bed, seen in both pictures for the distribution after 3 seconds from experimental and model setup, disperses much faster on its way to the top in the model than in the experiments. In the experiments after 4 seconds there can still be recognized a small dark layer at the top while this is not visible in the corresponding picture from the model.

We stress that the phase transition parameters seem to have a decisive impact on the model behavior and see that his quite simple model already

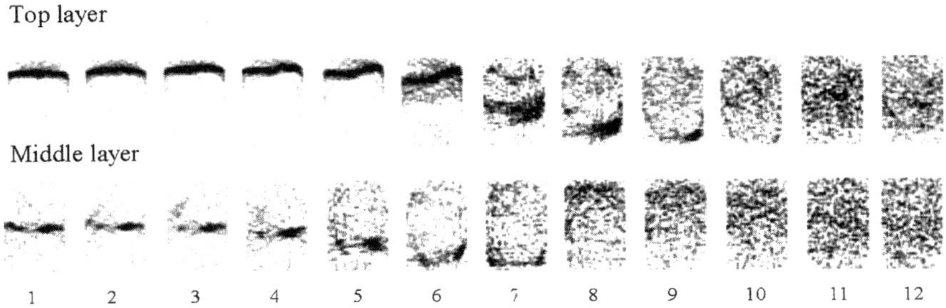

Figure 8.15: PET images showing the spread of pulses initially positioned at the top and in the middle of the fluidized bed. The time in seconds is given below the images. Details of the experimental and analysis method are given in [38]

captures the behavior of this complicated system for the most part. Quantifying the transition parameters, especially those for the phase transitions, is still an issue. Moreover Equation (8.4) has not yet been used in the setup of the model and should improve its quality when implemented correctly. As mentioned before the invariant distribution of such a system should be uniform. This can be recovered in the model when looking at Figure 8.16, too

Even when parameters cannot be calculated to represent a given system quantitatively the presented model still captures its qualitative features. A thorough study of an experiment as in [38] using the methods of this chapter should also give quantitatively correct results.

8.6 Review of Too et al.

Literature with the subject of multiphase systems understood as in this chapter is rare. Even a first step in the direction of a general and conceptual approach to modeling complex processes with multiphase systems has been missing in the literature and fortunately, can now be found in this chapter. Nevertheless some work has been done before in quite specific contexts. The article of Too et al. [119] stands out.

It features a gas-solids fluidized bed reactor with mixing and a chemical reaction taking place inside it. Three different phases describing a

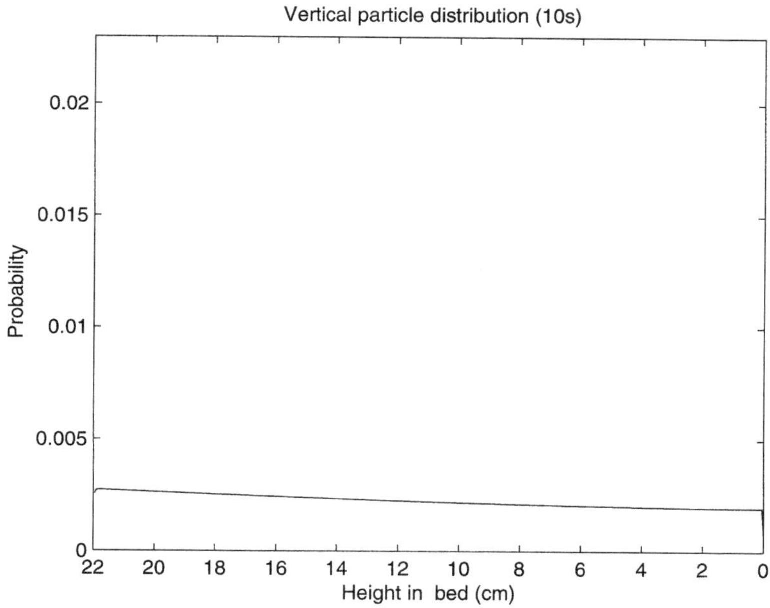

Figure 8.16: Stationary vertical particle distribution

molecule's location, its phase (dense or dilute) and its chemical type are modeled. The modeled fluidized bed reactor consists of ideally stirred tanks of various sizes. A time-heterogeneous system is set up. The time-dependent transition probabilities are calculated by using the so-called intensity matrix $K(t)$ which relates to the time-dependent transition matrix $P(s,t)$ via

$$\frac{d}{dt}P(s,t) = P(s,t)K(t) \qquad \text{and } P(t,t) = I \tag{8.6}$$

for all $t > s \geq 0$. The entries $p_{ij}(s,t)$ of $P(s,t)$ are equal to the probability of a molecule starting in cell i at time s to be in cell j at time t. The elements $k_{ij}(t)$ of the intensity matrix $K(t)$ are found as total volumetric flow from cell i to cell j divided by volume of cell i at time t. Given the intensity matrices $K(t)$ the transition matrices $P(s,t)$ are obtained as the solution of System (8.6).

Using a method outlined by Mori and Wen [93] the reactor is partitioned in ideally stirred tanks, so-called compartments. These compartments include dense and dilute molecules of all chemical types and are split in different states encoding chemical type and phase (dense or dilute). Although molecules are allowed to leave the reactor from the last compartment no exit cell is part of the state space. The volumetric flows between all compartments and inside each compartment are estimated using superficial gas velocities and material balance equations. With these parameters a time-independent intensity matrix is calculated, the corresponding transition matrix derived as solution of System (8.6) and some simulations are performed and discussed.

Too et al. formulate a heterogeneous Markov model in continuous time and discrete space (they later apply only a homogeneous one). Therefore they need to derive the intensity matrix to get hold of the transition matrix. The approach in the present chapter tries to use the transition matrix directly without detouring to the intensity matrix first and subsequently discretizes time as well. From the beginning a homogeneous case is assumed to simplify the presentation and increase the applicability. The discretization of space is a matter of choice of fineness and not specified by an underlying external theory as in the article by Too et al. where results from Mori and Wen [93] determine the compartment heights, i.e. the discretization fineness. Considerations of mass balances inside the reactor and between different cells are found in the article by Too et al., when parameters are estimated and in the present chapter on the level of restraints for the transition probabilities to be applied generally.

The multiphase Markov chain model of Too et al. for the gas-solids fluidized bed reactor with mixing and a chemical reactions constitutes a successful use of stochastic modeling particularly with a multiphase model. Too et al. point out that a complex system could be modeled with reduced efforts for formulating and solving the model equations and performing computational and manipulative tasks. Moreover the relation to a deterministic system described by a partial differential equation if the number of molecules is large and the extra insight in the microscopic system if not are discussed.

8.7 Danckwerts' law for a Multiphase Systems

As announced before we include a consideration of Danckwerts' law in case of the multiphase system. We study as an example for the general case the bubbling fluidized bed reactor presented in Section 8.1. Returning to the basic results for Danckwerts' law to hold from chapter 4 we find

$$
\begin{array}{llll}
1. & \sum_{j=1}^{N} p_{j,1} & = & 1 - v, \\
2. & \sum_{j=1}^{N} p_{j,i} & = & 1 \qquad \text{for all } 2 \leq i \leq N, \\
3. & \sum_{j=1}^{N} p_{j,N+1} & = & v,
\end{array}
$$

as necessary conditions on the transfer probabilities. At the first glance these conditions only apply to a state space of the form $\{1, 2, \ldots, N+1\}$. Nevertheless they can be used in the present case of the multiphase system for the bubbling fluidized bed reactor with its product state space $S = \{1, 2, \ldots, N\} \times \{0, 1\} \cup \{(N+1, 0)\}$. We just have to consider the states of S numbered from 1 to $2N+1$ and identify the entrances 1 and $(1, 0)$ and the exits $N+1$ and $(N+1, 0)$. This leads to the following conditions

$$
\begin{array}{llll}
(i) & \lambda_1^{(1)} + \delta_2^{(0)} + v & = & \beta_1^{(0)} + \lambda_1^{(0)}, \\
(ii) & \delta_2^{(1)} + \lambda_1^{(0)} & = & \lambda_1^{(1)}, \\
(iii) & \beta_{i-1}^{(0)} + \delta_{i+1}^{(0)} & = & \delta_i^{(0)} + \beta_i^{(0)} + \lambda_i^{(0)} \qquad \text{for all } 2 \leq i \leq N, \\
(iv) & \delta_{i+1}^{(1)} + \lambda_i^{(0)} & = & \delta_i^{(1)} \qquad \text{for all } 2 \leq i \leq N
\end{array}
$$

after calculating the inflows in each cell as follows

- for cell $(1, 0)$ we obtain $\lambda_1^{(1)} + 1 - (\beta_1^{(0)} + \lambda_1^{(0)}) + \delta_2^{(0)}$ as inflow,
- for cell $(1, 1)$ we obtain $\delta_2^{(1)} + \lambda_1^{(0)} + 1 - \lambda_1^{(1)}$ as inflow,
- for cells $(i, 0)$ the inflow is given as $\beta_{i-1}^{(0)} + \delta_{i+1}^{(0)} + 1 - (\delta_i^{(0)} + \beta_i^{(0)} + \lambda_i^{(0)})$ $(2 \leq i \leq N)$,
- for cells $(i, 1)$ the inflow is given as $\delta_{i+1}^{(1)} + 1 - \delta_i^{(1)} + \lambda_i^{(0)}$ $(2 \leq i \leq N)$

and equating via conditions 1.-3. above. The conditions (i)–(iv) above show that for all cells in the interior the inflow equals the outflow, i.e. mass is preserved and the net-outflow from the exit equals v.

In accordance to the considerations in sections 8.4 and 8.5 concerning the mass balance equation (8.4) it has to be taken into account that introducing several phases 'splits' one physical cell into several ones in the model. Therefore in might be necessary to associate the volumes of the original cell and the new ones in the model with each other in a convenient way and reflect this in the transition parameters. At the moment this is still an open problem but research to solve it is already in progress.

8.8 The abstract Multiphase System

Finally an abstract overview of the use of multiphase systems shall be given. As in the usual context of stochastic modeling the experimenter first has to identify the driving forces, effects and circumstances of the experiment which are necessary to hold onto when modeling. Care has to be taken that the model is still feasible and sensible maintaining the character of the original experiment. Indicators which suggest the use of a multiphase system are given if the evolution of the system $(X_n)_{n\geq 0}$ from one timestep n to the next $n+1$ ($X_n \rightsquigarrow X_{n+1}$) cannot be derived using one parameter, i.e. state, alone but a (finite) number of these or if the object of interest has different types of changing behaviors. Thereafter two crucial steps have to be taken. The state space and the transition probabilities must be specified.

The state space includes the information of interest, e.g. the particles location or the number of particles. It also has to reflect that the transition probabilities are only permitted to depend on the present state X_n and no other states in the past. Since the information of interest is in the case of a multiphase system not sufficient anymore to formulate a Markov chain the state space has to be enlarged beginning from the information of interest until the process is Markovian. These additional states encode the different phases whose descriptions have been detected to be essential when formulating the model in the first step.

Subsequently the transition probabilities have to be assigned. They determine the behavior of the system and need to continue the implementation of experimental conditions in the modeling context. It has to be ensured that the transition probabilities are consistent, i.e. greater than 0 and summing up to 1. Particular attendance should be given to the transition probabilities concerning changes of phase because these do not exist in simpler models. For practical use quantification of the model parameters especially the transition probabilities which can be a difficult task is of uttermost importance. If possible this observation should affect the choice of the state space. Qualitative knowledge about the system should enter the choice of the transition parameters as well. One can think of (local) mass balances or (known) long time behavior, e.g. knowledge of the invariant measure.

Finally the theory to Markov chains can be applied with full force using preliminary results. The transition matrix and its iterates give predictions of the system after some timesteps when starting from an initial setup. Long time behavior of the system can be deduced by considering the eigenvalues and -vectors of the transition matrix. Residence time distributions are available when applying techniques from Chapters 2, 4 and 5. It should also not be forgotten that information about Markov chain models can be exactly computed via matrix manipulations which is usually significant faster and more convenient than running simulations.

Chapter 9

Diffusion Limits

Almost all the models that we have introduced in the previous chapters are discrete in space and time. This facilitates the modeling process as transition probabilities in discrete Markov chains are much easier to specify than transition kernels in continuous Markov processes. At the same time, discrete models allow computations via matrix algebra. However, from a philosophical point of view, discrete models are less appealing since we are mostly modeling continuous processes, after all. In this chapter we will show how a continuous model can be obtained as a limit of discrete models. As an example, we will focus on the transport model in a freely bubbling fluidized bed, introduced in section 5.2 [1], though our approach can in principle be applied to other models as well.

9.1 Fokker-Planck equation

In section 5.2 we have introduced a modified birth-death process as model for the transport of particles in a bubbling fluidized bed without baffles. We will now investigate the behavior of this process in the diffusion limit, i.e. when time and space discretization converge to zero. We will show that the limit process exists and derive its Fokker-Planck equation.

In our model X_t denotes the height of the particle in the reactor measured as distance from the top. We take the reactor height as unit of length, so that in our model the height is 1. Let $v, D, \lambda : [0,1] \to \mathbb{R}$ be continuous functions. The functions $v(x), D(x)$ specify the drift and diffusion part of

[1] As mentioned the particle transport processes taking place in bubbling fluidized beds are the result of discrete events, namely the movement of fluidization bubbles, but they were, in line with the research literature, reformulated as continuous processes, and bubbling fluidization is therefore a suitable example here

the process, whereas $\lambda(x)$ specifies the rate at which particles are being caught in the wake of rising fluidization bubbles. For a small time interval $[t, t+\Delta t]$, the probability that a particle presently located at height x gets caught in a wake and deposited at the top, equals $\lambda(x)\,\Delta t$.

In order to obtain a discrete approximation we subdivide the reactor into N horizontal cells of equal height

$$\Delta = \frac{1}{N},$$

see Figure 9.1. The cells are indexed from top to bottom by the indices $1, 2, \ldots, N$. Thus, if a particle is in the i-th cell, its distance from the top is in the interval $[(i-1)\Delta, i\Delta)$. In addition, we introduce the exit state $E = N+1$. When a particle is in this state, it has left the reactor. We let the time discretization be

$$\epsilon = \frac{\Delta^2}{2\,D_0},$$

where $D_0 = \sup_{0\le x\le 1} D(x)$. Our computations later on will show that $\epsilon \sim \Delta^2$ is necessary, if we want to obtain a diffusion limit. The specific choice of ϵ, however, is somewhat arbitrary.

We first define a discrete time Markov chain $(\tilde{X}_n^\Delta)_{n\in\mathbb{N}}$ on the state space $\{1, \ldots, N, N+1\}$ with transition probabilities

$$\begin{aligned}
p_{i,i-1}^\Delta &= \left[\frac{\epsilon}{\Delta^2} D(i\Delta) - \frac{\epsilon}{2\Delta} v(i\Delta)\right](1-\epsilon\lambda(i\Delta)) \\
p_{i,i}^\Delta &= \left[1 - 2\frac{\epsilon}{\Delta^2} D(i\Delta)\right](1-\epsilon\lambda(i\Delta)) \\
p_{i,i+1}^\Delta &= \left[\frac{\epsilon}{\Delta^2} D(i\Delta) + \frac{\epsilon}{2\Delta} v(i\Delta)\right](1-\epsilon\lambda(i\Delta)) \\
p_{i,1}^\Delta &= \epsilon\lambda(i\Delta), \qquad (9.1)
\end{aligned}$$

for $2 \le i \le N$. For $i = 2$, the first and the last transition probability are added so that we get $p_{2,1}^\Delta = [\frac{\epsilon}{\Delta^2} D(i\Delta) - \frac{\epsilon}{2\Delta} v(i\Delta)](1-\epsilon\lambda(i\Delta)) + \epsilon\lambda(i\Delta)$. For the boundaries, we assume reflection at the origin and absorption in the last cell, i.e. we set

$$\begin{aligned}
p_{1,1}^\Delta &= 1 - \left[\frac{\epsilon}{\Delta^2} D(\Delta) + \frac{\epsilon}{2\Delta} v(\Delta)\right](1-\epsilon\lambda(\Delta)) \\
p_{1,2}^\Delta &= \left[\frac{\epsilon}{\Delta^2} D(\Delta) + \frac{\epsilon}{2\Delta} v(\Delta)\right](1-\epsilon\lambda(\Delta)) \\
p_{N+1,N+1}^\Delta &= 1. \qquad (9.2)
\end{aligned}$$

In addition we specify the initial value $\tilde{X}_0^\Delta = 1$, expressing the fact that the particle under consideration enters the reactor at time 0 at the top.

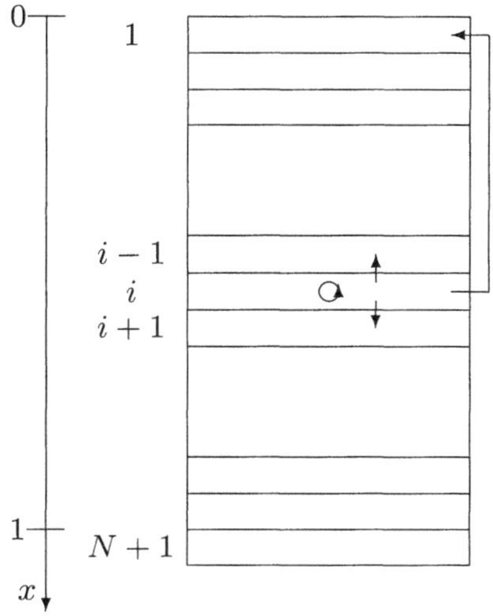

Figure 9.1: Transition probabilities for the discrete Markov chain $(\tilde{X}_n^{\Delta})_{n\in\mathbb{N}}$ and for the corresponding continuous time process $(X_t^{\Delta})_{t\geq 0}$, both with spatial discretization $\Delta = \frac{1}{N}$.

We then rescale the Markov chain $(\tilde{X}_n^{\Delta})_{n\in\mathbb{N}}$ and obtain the continuous time process

$$X_t^{\Delta} := \Delta \tilde{X}_{[t/\epsilon]}^{\Delta}, \quad t \geq 0,$$

on the state space $\{\Delta, 2\Delta, \ldots, 1-\Delta, 1, 1+\Delta\} \subset [0,1] \cup \{1+\Delta\}$. In what follows, we will investigate the asymptotic behavior of this process in the so-called diffusion limit, i.e. as $\Delta \to 0$ and thus also $\epsilon \to 0$. We fill first study the particle density and later the sample paths.

Let $\tilde{p}_{\Delta}(n,i)$ denote the probability that a particle is in cell i after n transitions, i.e.

$$\tilde{p}_{\Delta}(n,i) = P(\tilde{X}_n^{\Delta} = i).$$

If we assume that the particle distribution is uniform over the cell, we get at time t the particle density

$$p_{\Delta}(t,x) = \frac{1}{\Delta}\tilde{p}_{\Delta}([t/\epsilon], [x/\Delta]+1).$$

As $\Delta \to 0$, we expect $p_{\Delta}(t,x)$ to converge to the particle density $p(t,x)$ of the limit process $(X_t)_{t\geq 0}$; see Figure 9.2 for a numerical simulation. In the remaining part of this section, we will prove this assertion.

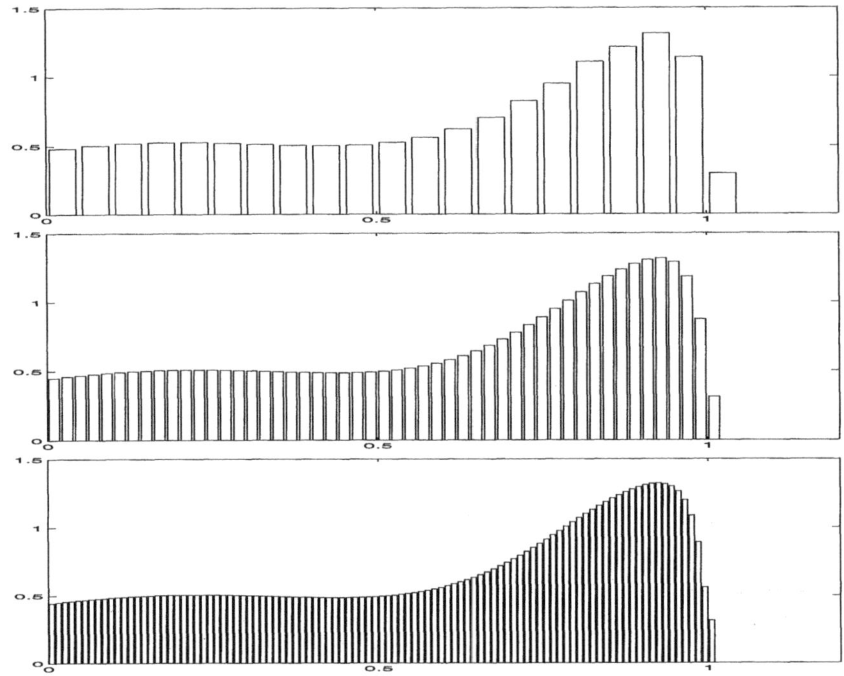

Figure 9.2: Particle density $p_\Delta(t,x)$ inside the reactor for three different discretizations, $\Delta = 1/20$, $\Delta = 1/50$ and $\Delta = 1/100$.

Another aspect of convergence in the diffusion limit concerns the residence time distribution (RTD). The cumulative RTD curve gives the probability that the particle has left the reactor by time t. In our model we get

$$\begin{aligned} F_\Delta(t) &= \tilde{p}_\Delta([t/\epsilon],[1/\Delta]+1) \\ &= 1-\sum_{i=1}^{N}\tilde{p}_\Delta([t/\epsilon],i) \\ &= 1-\int_0^1 p_\Delta(t,x)dx \to 1-\int_0^1 p(t,x)\,dx. \end{aligned}$$

Thus, as $\Delta \to 0$, $F_\Delta(t)$ converges to the cumulative RTD curve of the limit process $(X_t)_{t\geq 0}$, which is given by $F(t) = 1-\int_0^1 p(t,x)dx$.

We obtain the following Fokker-Planck equations for the approximate density $p_\Delta(t, x)$,

$$\begin{aligned}
&p_\Delta((n+1)\,\epsilon, i\,\Delta)\\
&= \left(\frac{\epsilon}{\Delta^2} D((i-1)\Delta) + \frac{\epsilon}{2\Delta} v((i-1)\Delta)\right)(1-\epsilon\,\lambda((i-1)\Delta))p_\Delta(n\,\epsilon,(i-1)\,\Delta)\\
&+ \left(1 - 2\frac{\epsilon}{\Delta^2} D(i\,\Delta)\right)(1 - \epsilon\,\lambda(i\,\Delta))p_\Delta(n\epsilon, i\Delta)\\
&+ \left(\frac{\epsilon}{\Delta^2} D((i+1)\Delta) - \frac{\epsilon}{2\Delta} v((i+1)\Delta)\right)(1-\epsilon\,\lambda((i+1)\Delta))p_\Delta(n\,\epsilon,(i+1)\,\Delta)\\
&= \left(\frac{\epsilon}{\Delta^2} D((i-1)\Delta) + \frac{\epsilon}{2\Delta} v((i-1)\Delta)\right)\\
&\quad \times (p_\Delta(n\epsilon,(i-1)\Delta)) - \epsilon\lambda((i-1)\Delta)p_\Delta(n\epsilon,(i-1)\Delta))\\
&+ \left(1 - 2\frac{\epsilon}{\Delta^2} D(i\Delta)\right)(p_\Delta(n\epsilon, i\Delta) - \epsilon\lambda(i\Delta)p_\Delta(n\epsilon, i\Delta))\\
&+ \left(\frac{\epsilon}{\Delta^2} D((i+1)\Delta) - \frac{\epsilon}{2\Delta} v((i+1)\Delta)\right)\\
&\quad \times (p_\Delta(n\epsilon, i\Delta) - \epsilon\lambda((i+1)\Delta)p_\Delta(n\epsilon,(i+1)\Delta))\\
&= \left(\frac{\epsilon}{\Delta^2} D((i-1)\Delta) + \frac{\epsilon}{2\Delta} v((i-1)\Delta)\right) p_\Delta(n\epsilon,(i-1)\Delta)\\
&+ \left(1 - 2\frac{\epsilon}{\Delta^2} D(i\,\Delta)\right) p_\Delta(n\epsilon, i\Delta)\\
&+ \left(\frac{\epsilon}{\Delta^2} D((i+1)\Delta) - \frac{\epsilon}{2\Delta} v((i+1)\Delta)\right) p_\Delta(n\epsilon, i\Delta)\\
&-\epsilon\lambda(i\Delta)p_\Delta(n\epsilon, i\Delta) + o(\epsilon).
\end{aligned}$$

valid for $i = 2, \ldots, N-1$. We then subtract $p(n\epsilon, i\Delta)$ from both sides, divide by ϵ, and get the partial difference equation

$$\begin{aligned}
&\frac{p_\Delta((n+1)\epsilon, i\Delta) - p_\Delta(n\epsilon, i\Delta)}{\epsilon}\\
&\quad = \frac{1}{\Delta^2}\{D((i-1)\Delta)p_\Delta(n\epsilon,(i-1)\Delta) - 2D(i\Delta)\,p_\Delta(n\epsilon, i\Delta)\\
&\qquad + D((i+1)\Delta)p_\Delta(n\epsilon,(i+1)\Delta)\}\\
&\qquad + \frac{1}{\Delta}\{v((i-1)\Delta)p_\Delta(n\epsilon,(i-1)\Delta) - v((i+1)\Delta)p_\Delta(n\epsilon,(i+1)\Delta)\}\\
&\qquad - \lambda(i\Delta)p_\Delta(n\epsilon, i\Delta) + o(1).
\end{aligned}$$

If we let $\Delta, \epsilon \to 0$ and at the same time let n, i depend on ϵ, Δ in such a way that $n\epsilon \to t$ and $i\Delta \to x$, the above difference equation approaches the partial differential equation

$$\frac{\partial}{\partial t}p(t,x) = \frac{\partial^2}{\partial x^2}(D(x)p(t,x)) - \frac{\partial}{\partial x}(v(x)p(t,x)) - \lambda(x)p(t,x). \quad (9.3)$$

Observe that this partial differential equation is obtained from the usual diffusion equation $\frac{\partial}{\partial t}p(t,x) = \frac{1}{2}\frac{\partial^2}{\partial x^2}(D(x)p(t,x)) - \frac{\partial}{\partial x}(v(x)p(t,x))$ by adding the extra term $-\lambda(x)p(t,x)$. The last term expresses the removal of mass as a result of wake transport to the top of the reactor.

We will next consider the boundary conditions at $x = 0$ and $x = 1$. At the entrance we obtain the following identity

$$\begin{aligned}
&p_\Delta((n+1)\epsilon, \Delta) \\
&= \epsilon\sum_{i=1}^{N}\lambda(i\Delta)p_\Delta(n\epsilon, i\Delta) \\
&\quad +(1-\epsilon\lambda(\Delta))\left(1 - \frac{\epsilon}{\Delta^2}D(\Delta) - \frac{\epsilon}{2\Delta}v(\Delta)\right)p_\Delta(n\epsilon,\Delta) \\
&\quad +(1-\epsilon\lambda(2\Delta))\left(\frac{\epsilon}{\Delta^2}D(2\Delta) - \frac{\epsilon}{2\Delta}v(2\Delta)\right)p_\Delta(n\epsilon,2\Delta) \\
&= \epsilon\sum_{i=1}^{N}\lambda(i\Delta)p_\Delta(n\epsilon, i\Delta) + \left(1 - \frac{\epsilon}{\Delta^2}D(\Delta) - \frac{\epsilon}{2\Delta}v(\Delta)\right)p_\Delta(n\epsilon,\Delta) \\
&\quad + \left(\frac{\epsilon}{\Delta^2}D(2\Delta) - \frac{\epsilon}{2\Delta}v(2\Delta)\right)p_\Delta(n\epsilon,2\Delta) + O(\epsilon).
\end{aligned}$$

We now subtract $p_\Delta(n\epsilon, \Delta)$ from both sides and then multiply both sides by Δ/ϵ. In this way we obtain

$$\begin{aligned}
&\Delta\frac{p_\Delta((n+1)\epsilon,\Delta) - p_\Delta(n\epsilon,\Delta)}{\epsilon} \\
&= \Delta\sum_{i=1}^{N}\lambda(i\Delta)p_\Delta(n\epsilon,i\Delta) + \frac{1}{\Delta}\left\{D(2\Delta)p_\Delta(n\epsilon,2\Delta) - D(\Delta)p_\Delta(n\epsilon,\Delta)\right\} \\
&\quad - \left\{\frac{1}{2}v(\Delta)p_\Delta(n\epsilon,\Delta) + \frac{1}{2}v(2\Delta)p_\Delta(n\epsilon,\Delta)\right\} + O(\Delta).
\end{aligned}$$

As we let $\Delta \to 0$, the left-hand side converges to zero. The first term on the right hand side is a Riemann sum and converges to $\int_0^1 \lambda(x)p(t,x)\,dx$. In the limit we thus get the following boundary condition at $x = 0$, i.e. at the entrance to the reactor,

$$\int_0^1 \lambda(x)p(t,x)dx + \frac{\partial}{\partial x}D(x)p(t,x)_{|x=0} - v(0)p(t,0) = 0. \quad (9.4)$$

Except for the $\int_0^1 \lambda(x)p(t,x)\,dx$ term, this is the usual reflecting boundary condition for a diffusion process. The extra term is a result of the deposition of mass at the top as a result of wake transport.

By similar computations we get the following boundary condition at the exit of the reactor, i.e. for $x = 1$,

$$\frac{\partial}{\partial x}(D(x)p(t,x))_{|x=1} = 0. \tag{9.5}$$

This equation expresses the fact that there is no flux through the exit due to diffusion. The only way in which particles can leave the reactor is via convection transport.

Note that the derivation of the PDE (9.3), together with the boundary conditions (9.4) and (9.5) is mainly heuristic. A rigorous proof of the fact that $p_\Delta(t,x)$ converges to a solution of this PDE can be found in Dehling, Hoffmann and Stuut ([39]).

9.2 Limit Process

In this section, we will give a probabilistic description of the stochastic process obtained by letting space and time discretization converge to zero. The essential idea lies in the decomposition of the Markov chain $(\tilde{X}_n^\Delta)_{n\in\mathbb{N}}$ into two components, namely a birth-death process and the process governing the jumps. By analyzing both components separately, we obtain a jump diffusion process as limit.

We first define a birth-death Markov chain $(\tilde{Y}_n^\Delta)_{n\in\mathbb{N}}$ on the state space $\{1,\ldots,N,N+1\}$ by $\tilde{Y}_0^\Delta = 1$ and the transition probabilities

$$\begin{aligned}
p_{i,i-1}^\Delta &= \left[\frac{\epsilon}{\Delta^2}D(i\Delta) - \frac{\epsilon}{2\Delta}v(i\Delta)\right]\\
p_{i,i}^\Delta &= \left[1 - 2\frac{\epsilon}{\Delta^2}D(i\Delta)\right]\\
p_{i,i+1}^\Delta &= \left[\frac{\epsilon}{\Delta^2}D(i\Delta) + \frac{\epsilon}{2\Delta}v(i\Delta)\right],
\end{aligned}$$

for $2 \leq i \leq N$, together with the boundary conditions

$$\begin{aligned}
p_{1,1}^\Delta &= 1 - \left[\frac{\epsilon}{\Delta^2}D(\Delta) + \frac{\epsilon}{2\Delta}v(\Delta)\right]\\
p_{1,2}^\Delta &= \left[\frac{\epsilon}{\Delta^2}D(\Delta) + \frac{\epsilon}{2\Delta}v(\Delta)\right]
\end{aligned}$$

at the entrance and $p_{N+1,N+1} = 1$ at the exit. Note that these are exactly the transition probabilities (9.1) and (9.2) with $\lambda(x) \equiv 0$.

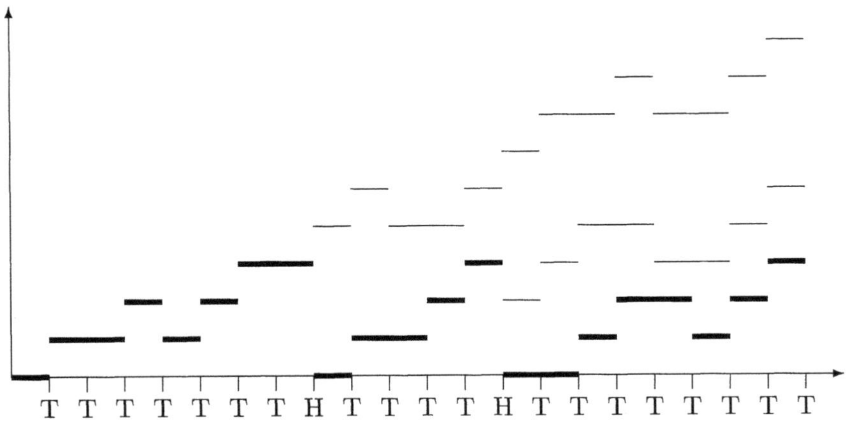

Figure 9.3: Two ingredients of birth-death process with jumps: ordinary birth-death process and process of coin tosses (H and T). The thick lines represent the final realization of the birth-death process with jumps.

We then rescale the Markov chain $(\tilde{Y}_n^{\Delta})_{n\in\mathbb{N}}$ in the same way as above and obtain the continuous time process

$$Y_t^{\Delta} := \Delta\, \tilde{Y}_{[t/\epsilon]}^{\Delta},\ t \geq 0.$$

As $\Delta \to 0$, by a well-known theorem on diffusion approximation, see e.g. Theorem 4.1. in Bhattacharya and Waymire ([12]), the process $(Y_t^{\Delta})_{t\geq 0}$ converges in distribution to a diffusion process $(Y_t)_{t\geq 0}$ with drift $v(x)$ and diffusion $D(x)$.

We now define a process of coin tosses associated with the rescaled birth-death chain $(Y_t^{\Delta})_{t\geq 0}$, and with probabilities for Head and Tail depending on the outcome of the process $(Y_t^{\Delta})_{t\geq 0}$. Coin tosses are performed at times $t_n = n\epsilon$, $n = 1, 2, \ldots$, with probabilities $\epsilon\, \lambda(Y_{t_n}^{\Delta})$ for Head and $1 - \epsilon\, \lambda(Y_{t_n}^{\Delta})$ for Tail. We stop the process as soon as Head occurs for the first time and then restart the process at 1, continuing this ad infinitum. We denote by

τ_Δ the time of the first Head, and find that

$$\begin{aligned} P(\tau^\Delta \geq t) &= P(\text{“Tail” at times } \epsilon, 2\epsilon, \ldots, [t/\epsilon]\,\epsilon) \\ &= \prod_{k=1}^{[t/\epsilon]} (1 - \epsilon\,\lambda(Y_{k\epsilon}^\Delta)) \\ &= \exp\left(\sum_{k=1}^{[t/\epsilon]} \log(1 - \epsilon\,\lambda(Y_{k\epsilon}^\Delta))\right) \approx \exp\left(-\sum_{k=1}^{[t/\epsilon]} \epsilon\,\lambda(Y_{k\epsilon}^\Delta)\right), \end{aligned}$$

where we have applied Taylor expansion of the logarithm in the last step. On the right hand side we find the Riemann sum $\epsilon \sum_{k=1}^{[t/\epsilon]} \lambda(Y_{k\epsilon}^\Delta)$, approximating the integral $\int_0^t \lambda(Y_s)ds$. Thus we may conclude

$$P(\tau^\Delta \geq t) \approx \exp\left(-\int_0^t \lambda(\tilde{Y}_{[s/\epsilon]}^\Delta)ds\right).$$

We now use the fact that the process $(\Delta\, Y_{[s/\epsilon]}^\Delta)_{s\geq 0}$ converges to the diffusion process $(Y_s)_{s\geq 0}$ to get $P(\tau^\Delta \geq t) \to \exp(-\int_0^t \lambda(Y_s)\,ds)$, as $\Delta \to 0$, and thus

$$P(\tau^\Delta \leq t) \to 1 - \exp\left(-\int_0^t \lambda(Y_s)\,ds\right).$$

The limit on the right hand side is the distribution function of the time τ until a first event in an inhomogeneous Poisson process with rate function $\lambda(Y_s)$. In total we may conclude that the stopping times τ^Δ converge in distribution to a stopping time τ, with this distribution.

We can then put the two ingredients together and show that the original rescaled process $(X_t^\Delta)_{t\geq 0}$ converges as $\Delta \to 0$ to a process $(X_t)_{t\geq 0}$ that can be described as follows: first we let a diffusion process $(Y_t)_{t\geq 0}$ run with drift $v(x)$ and diffusion $D(x)$ and then we take an inhomogeneous Poisson process on $[0, \infty)$ with rate $\lambda(Y_t)$. We stop the process Y_t at the time of the first event in the Poisson process, reset the process value at time τ to zero and restart the process. The mathematical details of this proof can be found in a paper by Dabrowski and Dehling ([29]).

Appendix A

Equations for RTD in CSTR and DPF

We give an account of the derivations of the F-curves for CSTRs in series and for the DPF model, Equations (4.14) and (4.15), respectively. We follow roughly the method used in [6], which has its origins in a treatment by Prof. H. Kramers of the Technical University of Delft. We use the notion of the inflow stream undergoing a step change from white to red material at $t = 0$.

A.1 Ideally Mixed Vessels (CSTRs) in Series

In a CSTR, the composition of the vessel content is always equal to the outlet composition.

To derive an expression for the the F-curve in one CSTR first, we perform a balance of red material over the CSTR:

$$\text{rate of accumulation} = \text{flow in} - \text{flow out}$$
$$\frac{dF(t)}{dt} = Q(1 - F(t)). \tag{A.1}$$

Changing to the dimensionless time $\theta := tV/Q$ and simplifying:

$$\frac{dF(\theta)}{d\theta} = 1 - F(\theta). \tag{A.2}$$

This is variables separable, and can be integrated directly between limits $(\theta = 0, F(0) = 0)$ and an arbitrary time $(\theta; F(\theta))$ to give:

$$F(\theta) = 1 - e^{-\theta}. \tag{A.3}$$

For a system consisting of multiple CSTRs of equal volume V/n in series, we denote the fraction of red material in the ith vessel by F_i (see Figure A.1), and that of the final, n^{th} , vessel with F.

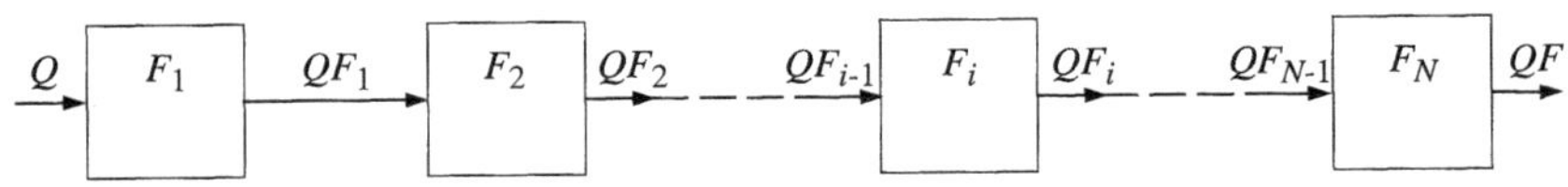

Figure A.1: A series of CSTRs of equal volume

A balance of red material for the ith vessel gives, after changing to dimensionless time:

$$\frac{dF(\theta)}{d\theta} + nF_i(\theta) - nF_{i-1}(\theta) = 0. \tag{A.4}$$

This is a first order, inhomogeneous equation, the solution of which can be written (dropping the argument of F):

$$F_i = e^{-n\theta} \int_0^\theta nF_{i-1} e^{n\theta} d\theta, \tag{A.5}$$

which can be verified by substitution. Substituting the variable of integration with $z := exp(n\theta)$ gives a recursion formula for F_i:

$$F_i = \frac{1}{z} \int_1^z F_{i-1} dz, \tag{A.6}$$

For $i = 1$, $F_{i-1} = 1$, so that:

$$F_1 = F = \frac{1}{z} \int_1^z dz = \frac{1}{z}(z-1) = 1 - e^{-\theta}, \tag{A.7}$$

the result we found above. For two mixers:

$$F_2 = F = \frac{1}{z} \int_1^z F_1 dz = \frac{1}{z} \int_1^z \frac{1}{z}(z-1)dz = \frac{1}{z}(z - 1 - \ln z). \tag{A.8}$$

Continuing this, we get for n mixers:

$$\begin{aligned} F_n = F &= \frac{1}{z}(z - 1 - \ln z - \frac{1}{2}(\ln z)^2 - \ldots - \frac{1}{(n-1)!}(\ln z)^{n-1}) \\ &= e^{-n\theta}(e^{n\theta} - 1 - n\theta - \frac{1}{2}(n\theta)^2 - \ldots - \frac{1}{(n-1)!}(n\theta)^{n-1}) \\ &= 1 - e^{-n\theta} \sum_{i=1}^{n} \frac{(n\theta)^{i-1}}{(i-1)!} \end{aligned} \tag{A.9}$$

A.2 Plug Flow with Axial Dispersion

To find F-curves for this RTD-model, we need to solve the one-dimensional convection-diffusion equation, 1.10:

$$\frac{\partial C(x,t)}{\partial t} = -v_x \frac{\partial C(x,t)}{\partial x} + D\frac{\partial^2 C(x,t)}{\partial x^2}. \tag{A.10}$$

with the appropriate boundary conditions.

If we solve the equation in a semi-infinite domain only one boundary condition is needed, and the length of the tube does not influence the concentration profile of red material. In that case, the number of independent variables can be reduced by the standard method of dimensional analysis[1]. This shows that the concentration (fraction) of red material can be written as a function of only one dimensionless independent variable: $z := (x - vt)/\sqrt{4Dt}$. Substituting both of the independent variables x and t with z and rearranging gives:

$$\frac{\partial^2 C(z)}{\partial z^2} + 2z\frac{\partial C(z)}{\partial z} = 0. \tag{A.11}$$

The solution to this second-order ordinary differential equation is:

$$C(z) = \frac{2C_1}{\sqrt{\pi}} \int_0^z e^{-u^2} du + C_2 = C_1 \mathrm{erf}(z) + C_2, \tag{A.12}$$

where erf is the error function. To find the constants of integration we use the following boundary conditions:

$$\begin{aligned} &\text{for } x \to \infty; z \to \infty \text{ and } C \to 0 \\ &\text{for } t \to \infty; z \to -\infty \text{ and } C \to 1. \end{aligned} \tag{A.13}$$

Now $\mathrm{erf}(\infty) = 1$ and $\mathrm{erf}(-\infty) = -1$, giving $C_1 = -1/2$ and $C_2 = 1/2$, so that the final solution for the concentration profile is:

$$C(z) = -\frac{1}{2}\mathrm{erf}(z) + \frac{1}{2} = -\frac{1}{2}\mathrm{erf}\left(\frac{x - vt}{\sqrt{4Dt}}\right) + \frac{1}{2}. \tag{A.14}$$

If we choose some position in the tube $x = L$ as the exit boundary and introduce the dimensionless time $\theta := tV/Q = tL = v$, basing the analysis on unit cross-sectional area, and the dimensionless parameter $Pe := vL/D$, called the *Peclet number*, we obtain for F:

$$F = -\frac{1}{2}\mathrm{erf}\left(\frac{L - vt}{\sqrt{4Dt}}\right) + \frac{1}{2} = -\frac{1}{2}\mathrm{erf}\left(\frac{1-\theta}{\sqrt{\frac{4}{Pe}\theta}}\right) + \frac{1}{2}. \tag{A.15}$$

This completes our derivation of these two very popular RTD models.

Bibliography

[1] J. C. Abanades and S. Atarés. Investigation of solid mixing in a deep fluidized bed of coarse particles by image analysis. In *Proceedings of the World Conference on Particle Technology 3*, number 211, July 1998.

[2] J. C. Abanades, S. Kelly, and G. P. Reed. A mathematical model for segregation of limestone-coal mixtures in slugging fluidized beds. *Chem. Eng. Sci.*, 43:3943–3953, 1994.

[3] J. Baeyens and D. Geldart. An investigation into slugging fluidized beds. *Chem. Eng. Sci.*, 29:255–265, 1974.

[4] T. Baron, C. L. Briens, P. Galtier, and M. A. Bergougnou. Verification of models and correlations for bubble properties in fluidized-beds. *Chem. Eng. Sci.*, 45:2227–2233, 1990.

[5] G. K. Batchelor. *An Introduction to Fluid Dynamics.* Cambridge University Press, 2000.

[6] W. J. Beek, K. M. K. Muttzall, and J. W. van Heuven. *Transport Phenomena.* John Wiley & Sons, second edition, 1999.

[7] F. Berruti, A. G. Liden, and D. S. Scott. Measuring and modelling residence time distribution of low density solids in a fluidized bed reactor of sand particles. *Chem. Eng. Sci.*, 43:739–748, 1988.

[8] H. Berthiaux. Analysis of grinding processes by markov chains. *Chem. Eng. Sci.*, 55:4117–4127, 2000.

[9] H. Berthiaux and J. Dodds. A new estimation technique for the determination of breakage and selection parameters in batch grinding. *Powder Technol.*, 94:173–179, 1997.

[10] F. Bezzo and S. Macchietto. A general methodology for hybrid multizonal/CFD models. Part II. Automatic zoning. *Computers and Chemical Engineering*, 28:513–525, 2004.

[11] F. Bezzo, S. Macchietto, and C. C. Pantelides. A general methodology for hybrid multizonal/CFD models. Part I. Theoretical framework. *Computers and Chemical Engineering*, 28:501–511, 2004.

[12] R. N. Bhattacharya and E. C. Waymire. *Stochastic Processes with Applications*. John Wiley and Sons, 2000.

[13] H. T. Bi, J. Zhou, S.-Z. Qin, and J. R. Grace. Annular wall layer thickness in circulating fluidized bed risers. *Can. J. Chem. Eng.*, 74:811–814, 1996.

[14] C. A. Biggs, C. Sanders, A. C. Scott, A. W. Willemse, A. C. Hoffmann, T. Instone, A. D. Salman, and M. J. Hounslow. Coupling granule properties and granulation rates in high-shear granulation. *Powder Technol.*, 130:162–168, 2003.

[15] E. Bilgili and B. Scarlett. Nonlinear effects in particulate processes. *Nonlinear Analysis*, 63:e1131–e1141, 2004.

[16] P. Billingsley. Prime numbers and Brownian motion. *The American Mathematical Monthly*, 80:1099–1115, 1973.

[17] R. B. Bird, W. E. Stewart, and E. N. Lightfoot. *Transport Phenomena*. John Wiley and Sons, second edition, 2002.

[18] G. A. Bokkers, M. van Sint Annaland, and J. A. M. Kuipers. Mixing and segregation in a bidispers gas–solid fluidised bed: a numerical and experimental study. *Powder Technology*, in press, 2004.

[19] J. C. Bosma and A. C. Hoffmann. On the capacity of continuous powder classification in a gas-fluidized bed with horizontal sieve-like baffles. *Powder Technology*, 134:1–15, 2003.

[20] P. Bradshaw. *An Introduction to Turbulence at its Measurement*. Pergamon Press, 1975.

[21] H. S. Carslaw and J. C. Jaeger. *The Conduction of Heat in Solids*. Oxford University Press, second edition, 1986.

[22] S. G. Chapman and T. G. Cowling. *The Mathematical Theory of Nonuniform Gases*. Cambridge University Press, 1961.

[23] L. Cheung, A. W. Nienow, and P. N. Rowe. Minimum fluidization velocity of a binary mixture of different sized particles. *Chem. Eng. Sci.*, 29:1301–1303, 1974.

[24] P. W. Cleary and M. L. Sawley. Dem modelling of industrial granular flows: 3d case studies and the effect of particle shape on hopper discharge. *Applied Mathematical Modelling*, 26:89–111, 2002.

[25] R. Clift, J. R. Grace, and M. E. Weber. *Bubbles, Drops and Particles.* Dover Publications, 2005.

[26] J. M. Coulson and J. F. Richardson. *Chemical Engineering*, volume 2. Pergamon Press, third edition, 1980.

[27] C. Crowe, M. Sommerfeld, and Y. Tsuji. *Multiphase Flow with Droplets and Particles.* CRC Press, 1998.

[28] Y. Q. Cui, R. G. J. M. van der Laans, H. J. Noordman, and K. Ch. A. M. Luyben. Compartment mixing model for stirred reactors with multipel impellers. *Trans IChemE*, 74:261–271, 1996.

[29] A. R. Dabrowski and H. G. Dehling. Jump diffusion approximation for a markovian transport model. In *Asymptotic methods in probability and statistics*, pages 115–125. Elsevier Science, Amsterdam, 1998.

[30] P. V. Danckwerts. Continuous flow systems. distribution of residence times. *Chem. Eng. Sci.*, 2:1–13, 1953.

[31] J. F. Davidson, R. Clift, and D. Harrison. *Fluidization.* Academic Press, New York, second edition, 1985.

[32] J. F. Davidson and D. Harrison. *Fluidization.* Academic Press, New York, 1971. chapter 5.

[33] C. Dechsiri. *Particle transport in fluidized beds; experiments and stochastic models.* PhD thesis, University of Groningen, Dept. of Chemical Engineering, 2004.

[34] C. Dechsiri, J. C. Abanades, H. G. Dehling, A. C. Hoffmann, and P. C. Wright. A stochastic model for mixing and segregation in slugging fluidized beds. In *Proceedings of the Third European Conference on Fluidization*, pages 631–638, 2000.

[35] C. Dechsiri, J. C. Bosma, A. C. Hoffmann, G. Hui, and H. G. Dehling. A stochastic model for particle mixing and segregation in fluidized bed with baffles. In *Proceedings PARTEC 2001, International Congress for Particle Technology*, Nünberg, March 2001.

[36] C. Dechsiri, H. G. Dehling, and A. C. Hoffmann. Stochastic model for fluidization. In *Proceedings of the 4th World Congress on Particle Technology*, 2002.

[37] C. Dechsiri, A. Ghione, F. van der Wiel, H. G. Dehling, A. M. J. Paans, and A. C. Hoffmann. Positron emission tomography applied to fluidization engineering. *Can. J. Chem. Eng.*, 83:88–96, 2005.

[38] C. Dechsiri, E. A. van der Zwan, H. G. Dehling, and A. C. Hoffmann. The dispersion of particle pulses in fluidized beds measured by PET. *AIChE Journal*, 51:791–801, 2005.

[39] H. G. Dehling, H. W. Stuut, and A. C. Hoffmann. Stochastic models for transport in a fluidized bed. *SIAM J. Appl. Math.*, 60:337–358, 1999.

[40] J. J. Derksen. Separation performance predictions of a Stairmand high-efficiency cyclone. *AIChE Journal*, 49:1359–1371, 2003.

[41] J. J. Derksen. Simulations of confined turbulent vortex flow. *Computers & Fluids*, 34:301–318, 2005.

[42] A. Einstein. Investigations on the theory of the brownian movement. In R. Fürth, editor, .. Dover Publ., 1956.

[43] L. T. Fan, B. C. Shen, and S. T. Chou. Stochastic modeling of transient residence–time distributions during start–up. *Chem. Eng. Sci.*, 50:211–221, 1995.

[44] L. T. Fan, J. R. Too, and R. Nassar. Stochastic simulation of residence time distribution curves. *Chem. Eng. Sci.*, 40:1743–1749, 1985.

[45] W. Feller. Zur theorie der stochastischen prozesse. *Mathematische Annalen*, 113:113–160, 1936.

[46] D. Geldart. Effect of particle-size and size distribution on behaviour of gas-fluidized beds. *Powder Technology*, 6:201–215, 1972.

[47] L. G. Gibilaro. Mean residence times in continuous flow systems. *Nature*, 270:47–48, 1977.

[48] L. G. Gibilaro. Residence time distributions in regions of continuous flow systems. *Chem. Eng. Sci.*, 34:697–702, 1979.

[49] L. G. Gibilaro, H. W. Kropholler, and D. J. Spikins. Solutions of a mixing model due to van de vusse by a simple probability method. *Chem. Eng. Sci.*, 22:517–523, 1967.

[50] L. G. Gibilaro and P. N. Rowe. A model for a segragating gas fluidised bed. *Chem. Eng. Sci.*, 29:1403–1412, 1974.

[51] W. J. Golz and J. R. Dorroh. The convection–diffusion equation for a finite domain with time varying boundaries. *Applied Mathematics Letters*, 14:983–988, 2001.

[52] T. Gottschalk, H. G. Dehling, and A. C. Hoffmann. Danckwerts' law for mean residence time revisited. *Chem. Eng. Sci.*, 61:6213–6217, 2006.

[53] A. K. Haines, R. P. King, and E. T. Woodburn. The interrelationship between bubble motion and solids mixing in a gas fluidized bed. *AIChE Journal*, 18:591–599, 1972.

[54] J. Happel and H. Brenner. *Low Reynolds Number Hydrodynamics: With Special Applications to Particulate Media.* Kluwer Print on Demand, 1983.

[55] A. T. Harris, J. F. Davidson, and R. B. Thorpe. Particle residence time distributions in circulating fluidised beds. *Chem. Eng. Sci.*, 58:2181–2202, 2003.

[56] A. T. Harris, R. B. Thorpe, and J. F. Davidson. Stochastic modelling of the particle residence time distribution in circulating fluidised bed risers. *Chem. Eng. Sci.*, 57:4779–4796, 2002.

[57] G. J. Hartholt. *Particle mixing in gas fluidized beds.* PhD thesis, University of Groningen, The Netherlands, 1996.

[58] G. P. Hartholt, A. C. Hoffmann, and L. P. B. M. Janssen. Visual observation of individual particle behaviour in gas and liquid fluidized beds. *Powder Technology*, 88:341–345, 1996.

[59] P. M. Heertjes, L. H. de Nie, and J. Verloop. Transport and residence time of particles in a shallow fluidized bed. In A. A. H. Drinkenburg, editor, *Proceedings of the International Symposium on Fluidization*, Eindhoven, the Netherlands, 1967. Netherlands University Press.

[60] J. O. Hinze. *Turbulence.* McGraw–Hill, 1975.

[61] A. C. Hoffmann, C. Dechsiri, F. van der Wiel, and H. G. Dehling. PET investigation of a fluidized particle: spatial and temporal resolution and short term motion. *Meas. Sci. Technol.*, 16:851–858, 2005.

[62] A. C. Hoffmann and H. G. Dehling. A stochastic modeling approach to particle residence time distribution in continuous fluidized beds. In *Proceedings of the World Conference on Particle Technology 3*, number 258, July 1998.

[63] A. C. Hoffmann, L. P. B. M. Janssen, and J. Prins. Particle segregation in fluidized binary mixtures. *Chem. Eng. Sci.*, 48:1583–1592, 1993.

[64] A. C. Hoffmann and H. Paarhuis. A study of the particle residence time distribution in continuous fluidized beds. *I. Chem. E. Symp. Ser.*, 121:37–49, 1990.

[65] M. J. Hounslow, R. L. Lyall, and V. R. Marshall. A discretized population balance for nucleation, growth and aggregation. *AIChE Journal*, 34:1821–1832, 1988.

[66] M. J. Hounslow, J. M. K. Pearson, and T. Instone. Tracer studies of high-shear granulation: II. Population balance modeling. *AIChE Journal*, 47:1984–1999, 2001.

[67] H. Hristov, R. Mann, V. Lossev, S. D. Vlaev, and P. Seichter. A 3-d analysis of gas-liquid mixing, mass transfer and bioreaction in a stirred bio-reactor. *Food and Bioproducts Processing*, 79:232–241, 2001.

[68] H. V. Hristov, R. Mann, V. Lossev, and S. D. Vlaev. A simplified cfd for three-dimensional analysis of fluid mixing, mass transfer and bioreaction in a fermenter equipped with triple novel geometry impellers. *Food and Bioproducts Processing*, 82:21–34, 2004.

[69] K. Îto. Differential equations determining a markov process. *J. Pan-Japan Math. Coll.*, 1077, 1942.

[70] K. D. Kafui, C. Thornton, and M. J. Adams. Discrete particle–continuum fluid modelling of gas–solid fluidized beds. *Chem. Eng. Sci.*, 57:2395–2410, 2002.

[71] M. Kagoshima and R. Mann. Development of a networks-of-zones fluid mixing model for an unbaffled stirred vessel used for precipitation. *Chem. Eng. Sci.*, 61:2852–2863, 2006.

[72] S. Karlin and H. G. Taylor. *A First Course in Stochastic Processes.* Academic Press, 1975.

[73] S. Karlin and H. G. Taylor. *A Second Course in Stochastic Processes.* Academic Press, 1981.

[74] K. Kato and C. Y. Wen. Bubble assemblage model for fluidized bed catalytic reactors. *Chem. Eng. Sci.*, 24:1351–1369, 1969.

[75] M. G. Kendall. *Time-Series.* Charles Griffin & Co. ltd., London, 2nd edition, 1985.

[76] P. E. Kloeden and E. Platen. *Numerical Solution of Stochastic Differential Equations.* Springer Verlad, 1992.

[77] E. Klose and W. Heschel. Zur messung der partikelverweilzeitverteilung in wirbelschichten mit gas-feststoff-reaktion. *Chem. Techn.*, 37:149–152, 1985.

[78] A. Kolmogorov. Über die analytischen methoden in der wahrscheinlichkeitsrechnung. *Mathematische Annalen*, 104:415–458, 1931.

[79] E. Kreyszig. *Advanced Engineering Mathematics.* John Wiley and Sons, sixth edition, 2006.

[80] K. Krishnaiah, Y. Pydisetty, and Y. B. G. Varma. Residence time distribution of solids in multistage fluidization. *Chem. Eng. Sci.*, 37:1371–1377, 1982.

[81] R. la Rivière, G. P. Hartholt, A. C. Hoffmann, and L. P. B. M. Janssen. Methods for the determination of particle residence time distribution in continuous gas fluidized beds. *I. Chem. E. Symp. Ser.*, 140:283–294, 1996.

[82] H. Lamb. *Hydrodynamics.* Dover Publ., sixth edition, 1986.

[83] O. Levenspiel. *Chemical Reaction Engineering.* John Wiley & Sons, third edition, 1999.

[84] C.-C. Lin and L. A. Segel. *Mathematics Applied to Deterministic Problems in the Natural Sciences.* SIAM, 1988.

[85] R. Mann. Gas-liquid stirred vessel mixers—towards a unified theory based on networks-of-zones. *Chem. Eng. Res. & Des.*, 64:23–34, 1986.

[86] R. Mann, P. P. Mavros, and J. C. Middleton. A structured stochastic flow model for interpreting flow-follower data from a stirred vessel. *Trans. I. Chem. E.*, 59:271–278, 1981.

[87] R. Mann, R. A. Williams, T. Dyakovski, F. J. Dickin, and R. B. Edwards. Development of mixing models using electrical resistance tomography. *Chem. Eng. Sci.*, 82:2073–2085, 1997.

[88] L. Massimilla and S. Bracale. Il mescolamento della fase solida nei sistemi: Solido-gas fluidizzati, liberi e frenati. *La Ricerca Scientifica*, 27:1509–1526, 1957.

[89] J. M. Matsen. Scale-up of fluidized bed processes: principles and practice. *Powder Technol.*, 88:237–244, 1996.

[90] J. M. Matsen, S. Hovmand, and J. F. Davidson. Expansion of fluidized beds in slug flow. *Chem. Eng. Sci.*, 24:1743–1754, 1969.

[91] J. M. D. Merry and J. F. Davidson. Gulf-stream circulation in shallow fluidized-beds. *Trans. Instn. Chem. Engrs.*, 81:361–368, 1973.

[92] S. Mori and C. Y. Wen. Estimation of bubble diameter in gaseous fluidized-beds. *AIChE Journal*, 21:109–115, 1975.

[93] S. Mori and C. Y. Wen. Simulation of fluidized bed reactor performance by modified bubble assemblage model. In D. L. Keairns, editor, *Fluidization Technology.* Hemisphere, Washington, DC, 1976.

[94] D. R. Morris, K. E. Gubbins, and S. B. Watkins. Residence time studies in fluidized and moving beds with continuous solids flow. *Trans. Inst. Chem. Engrs.*, 42:T323–T333, 1964.

[95] R. Nassar, L.-T. Fan, J. R. Too, and L.-S. Fan. A stochastic treatment of unimolecular reactions in an unsteady state continuous flow system. *Chem. Eng. Sci.*, 36:1307–1317, 1981.

[96] R. Nassar, J. R. Too, and L. T. Fan. Stochastic modeling of polymerization in a continuous flow reactor. *J. Applied Polymer Science*, 26:3745–3759, 1981.

[97] R. Nassar, J. R. Too, and L. T. Fan. Stochastic diffusion model for crystal size distribution in an open flow system. *AIChE Journal*, 30:1014–1016, 1984.

[98] R. Nassar, J. R. Too, and L. T. Fan. A probabilistic model of the Fischer-Tropsch synthesis in a flow reactor. *Chem. Eng. Commun.*, 43:287–300, 1986.

[99] T. E. Nichols. *Spatiotemporal modeling of positron emission tomography.* PhD thesis, Carnegie Mellon University, Dept. of Statistics, 2001.

[100] S. V. Patankar. *Numerical heat transfer and fluid flow.* Hemisphere Publishing Corp., 1980.

[101] R. H. Perry, D. W. Green, and J. O. Maloney, editors. *Perry's Chemical Engineer's Handbook.* McGraw-Hill, seventh edition, 1995.

[102] W. H. Press, S. A. Teukolsky, W. T. Vetterling, and B. P. Flannery. *Fortran numerical recipes.* Cambridge University Press, 1992.

[103] F. Pudel, M. Strümke, and U. Sündermann. Untersuchung des partikelverweilzeitverhaltens in gas-feststoff-wirbelrinnen. *Wiss. Z. Tech. Hochsch. Magdeburg*, 30:51–53, 1986.

[104] X. Qi. Stochastic models for prodrug targeting. 1. diffusion of the efflux drug. *Molecular Pharmeceutics*, 3:187–195, 2006.

[105] D. Ramkrishna. *Population Balances, Theory and Applications to Particulate Systems in Engineering.* Academic Press, 2000.

[106] A. D. Randolph and M. A. Larson. *Theory of Particulate Processes.* Academic Press, 1988.

[107] P. N. Rowe. Prediction of bubble-size in a gas-fluidized bed. *Chem. Eng. Sci.*, 31:285–288, 1976.

[108] P. N. Rowe and B. A. Partridge. Particle movement caused by bubbles in a fluidised bed. In *Interaction Between Fluids & Particles*, pages 135–142. Instn Chem. Engrs, London, 1962.

[109] M. Rubinovitch and U. Mann. Single-particle approach for analyzing flow systems. Part I: Visits to flow regions. *AIChE Journal*, 24:658–662, 1983.

[110] M. Rubinovitch and U. Mann. Single-particle approach for analyzing flow systems. Part II: Regional residence times and local flow rates. *AIChE Journal*, 24:663–668, 1983.

[111] M. Rubinovitch and U. Mann. Single-particle approach for analyzing flow systems. Part III: Multiple fluids. *AIChE Journal*, 31:615–620, 1985.

[112] J. C. Schouten, P. J. M. Valkenburg, and C. M. van den Bleek. Segregation in a slugging fbc large-particle system. *Powder Technol.*, 54:85–98, 1988.

[113] A. C. Scott, M. J. Hounslow, and T. Instone. Direct evidence of heterogeneity in high-shear granulation. *Powder Technol.*, 113:205–213, 2000.

[114] S. J. Stanley, H. Hristov, R. Mann, and K. Primrose. Reconciling electrical resistance tomography (ert) measurements with a fluid mixing model for semi-batch operation of a stirred vessel. *Can. J. Chem. Eng.*, 83:48–54, 2005.

[115] P. Szymczak and A. J. C. Ladd. Boundary conditions for stochastic solutions of the convection–diffusion equation. *Physical Review E*, 68:036704–1–036704–12, 2003.

[116] H. Tanimoto, S. Chiba, T. Chiba, and H. Kobayashi. Jetsam descent induced by a singel bubble passage in 3-dimensional gas-fluidized beds. *Journal of Chemical Engineering of Japan*, 14:273–276, 1981.

[117] J. R. Too, L. T. Fan, and R. Nassar. Markov chain models of complex chemical reactions in continuous flow reactors. *Computers and Chemical Engineering*, 7:1–12, 1983.

[118] J. R. Too, L. T. Fan, and R. Nassar. A stochastic axial dispersion model for tubular flow reactors. *Chem. Eng. Sci.*, 41:2341–2346, 1986.

[119] J. R. Too, R. O. Fox, L. T. Fan, and R. Nassar. Stochastic modeling of a fluidized-bed reactor. *AIChE Journal*, 31:992–998, 1985.

[120] R. D. Toomey and H. F. Johnstone. Gaseous fluidization of solid particles. *Chem. Eng. Progress*, 48:220–226, 1952.

[121] A. P. Torres and F. A. R. Oliveira. Residence time distribution studies in continuous thermal processing of liquid foods: a review. *Journal of Food Engineering*, 36:1–30, 1998.

[122] G. Tripathi, G. N. Pandey, and P. C. Singh. Residence time distribution studies on different systems: Derivation of a generalized correlation. *Indian J. Techn.*, 9:281–284, 1971.

[123] J. G. van de Vusse. A new model for the stirred tank reactor. *Chem. Eng. Sci.*, 17:507–521, 1962.

[124] E. H. van der Meer, R. B. Thorpe, and J. F. Davidson. Flow patterns in the square cross-section riser of a circulating fluidised bed and the effect of riser exit design. *Chem. Eng. Sci.*, 55:4079–4099, 2000.

[125] J.-J. van Dijk, A. C. Hoffmann, D. Cheesman, and J. G. Yates. The influence of horizontal internals on the flowpattern in dense fluidized beds. *Powder Technology*, 98:273–278, 1998.

[126] J. Verloop, L. H. de Nie, and P. M. Heertjes. The residence time of solids in gas-fluidized beds. *Powder Techn.*, 2:32–42, 1968/69.

[127] H. K. Versteeg and W. a Malalaseker. *An introduction to computational fluid dynamics: the finite volume method.* Addison–Wesley Publishing Co., 1996.

[128] S. D. Vlaev, R. Mann, and V. Lossev. An analysis of the effect of rheology on local gas hold-up—the case of thylosin production. *Chem. Eng. Res. & Des.*, 73:320–324, 1995.

[129] K. S. Wang and M. J. Rhodes. Mechanistic study of defluidization by numerical simulation. *Chem. Eng. Sci.*, 59:215–222, 2004.

[130] B. Weber and K. Rose. Mathematische modellierung von wirbelschichten. *Chem. Techn.*, 22:594–596, 1970.

[131] F. Wei and J. Zhu. Effect of the flow direction on axial solid dispersion in gas-solid cocurrent upflow and downflow systems. *Chem. Eng. J.*, 16:109–113, 1993.

[132] C. Y. Wen and Y. H. Yu. A generalized method for predicitng the minimum fluidization velocity. *AIChE Journal*, 12:610–612, 1966.

[133] J. Werther. Influence of the bed diameter on the hydrodynamics of gas fluidized beds. *AIChE Symp. Ser.*, 141:53–62, 1973.

[134] J. Werther. Influence of the bed diameter on the hydrodynamics of gas fluidized beds. *AIChE Symp. Ser.*, 70:53–62, 1974.

[135] K. R. Westerterp, W. P. M. van Swaaij, and A. A. C. M. Beenackers. *Chemical Reactor Design and Operation.* John Wiley & Sons, 1988.

[136] K. Whittmann, D. Wippern, H. Schlingmann, H. Helmrich, and K. Schügerl. Solid particle mixing in a continuously operated fluidized bed reactor. *Chem. Eng. Sci.*, 38:1391–1397, 1983.

[137] N. Wiener. Differential space. *J. Math. and Phys.*, 2:131–174, 1923.

[138] W. C. Yang. A model for a dynamics of a circulating fluidised bed loop. In P. Basu and J. F. Large, editors, *Circulating fluidised bed technology*, pages 181–191. Pergamon Press, 1988.

[139] J. Zahradník, R. Mann, M. Fialová, D. Vlaev, S. D. Vlaev, V. Lossev, and P Seichter. A networks-of-zones analysis of mixing and mass transfer in three industrial bioreactors. *Chem. Eng. Sci.*, 56:485–492, 2001.

Index

Mathematics in Science and Engineering
Edited by C.K. Chui, Stanford University

Recent titles:
I. Podlubny, *Fractional Differential Equations*
E. Castillo, A. Iglesias, R. Ruíz-Cobo, *Functional Equations in Applied Sciences*
V. Hutson, J.S. Pym, M.J. Cloud, *Applications of Functional Analysis and Operator Theory (Second Edition)*
V. Lakshmikantham and S.K. Sen, *Computational Error and Complexity in Science and Engineering*
T.A. Burton, *Volterra Integral and Differential Equations (Second Edition)*
E.N. Chukwu, *A Mathematical Treatment of Economic Cooperation and Competition Among Nations: with Nigeria, USA, UK, China and Middle East Examples*
Wei-Bin Zhang, *Discrete Dynamical Systems, Bifurcations and Chaos in Economics*
C. De Coster and P. Habets, *Two-Point Boundary Value Problems: Lower and Upper Solutions*
V.V. Ivanov and N. Ivanova, *Mathematical Models of the Cell and Cell Associated Objects*
Z. Zong, *Information-Theoretic Methods for Estimating Complicated Probability Distributions*
A.G. Ramm, *Dynamical Systems Method for Solving Operator Equations*
J. Mishra and S.N. Mishra, *L-System Fractals*
I.V. Konnov, *Equilibrium Models and Variational Inequalities*
H.G. Dehling, T. Gottschalk and A.C. Hoffmann, *Stochastic Modelling in Process Technology*

www.ingramcontent.com/pod-product-compliance
Ingram Content Group UK Ltd.
Pitfield, Milton Keynes, MK11 3LW, UK
UKHW020246290926
15470UKWH00002B/773

9 780444 5202